GEOTHERMAL ENERGY TECHNOLOGY ASSESSMENT

Paul N. Cheremisinoff
Angelo C. Morresi

a TECHNOMIC® publication
TECHNOMIC Publishing Co., Inc.
265 W. State St., Westport, Conn. 06880

GEOTHERMAL ENERGY TECHNOLOGY ASSESSMENT

265 W. State St., Westport, CT 06880

a TECHNOMIC® publication

Printed in U.S.A.
Library of Congress Card No. 75-43009
I.S.B.N. 0-87762-185-3

ABOUT THE AUTHORS

Paul Cheremisinoff, is Associate Professor of environmental engineering at the New Jersey Institute of Technology, and a registered professional engineer. Author of numerous technical papers on environmental engineering and management, he is author-editor of several major engineering handbooks.

Mr. Cheremisinoff is a consulting engineer to private industry, Associate Editor of Pollution Engineering magazine, an international publication. He is a diplomate of the American Academy of Environmental Engineers; Fellow of the New York Academy of Sciences; Sigma Xi.

Angelo Morresi, since earning his Bachelor's degree in Mechanical Engineering, Magna Cum Laude, also obtained a Master of Science in Environmental Engineering from New Jersey Institute of Technology. Contributor to industrial press publications, he has authored sections in two engineering handbooks.

Mr. Morresi is presently an engineer with Exxon Research & Engineering Co., having previously been associated in engineering capacities with Wallace & Tiernan Div. of Pennwalt Corp.

PREFACE

Geothermal energy – a vast reservoir bottled up beneath the earth's crust like a great genie to be uncorked. Can Geogenie be harnessed to meet man's energy requirements? Geothermal energy activities are progressing on a worldwide basis. Unfortunately so far practical results in usable energy are sparse in relation to societal needs.

Geothermal energy as the authors show has been proven locally significant in a few places. The Geogenie faces obstacles, some of which are formidable before a worldwide significant energy role can be achieved. The range of historical activity nonetheless is impressive even though current exploitation is somewhat lacking and energy of course is the big prize.

Energy, the driving force of the world has been fueled by fossil sources that has historically released man from the constraints of subsistence agriculture and home industry. World consumption of energy from fossil fuels has increased at approximately 2% per year. The United States is the world's leader in not only per capita income, but per capita energy consumption, having 6% of the world's population while consuming one-third of the world's energy production.

Unlimited supplies of low-cost energy have long been taken for granted in our country and all of our long range projections for increases in gross national product; improvements in life quality have been made based on the assumption of a continuous abundance of cheap energy. However, it is now become obvious that we have fast fallen from energy self-sufficiency into dependence on foreign sources to meet these needed requirements.

The problems associated with energy needs are extremely complex. It has however become apparent that the United States must develop all the energy supplies possible with present and future technologies. Among the sources of energy for likely candidates for development in substantial quantities is geothermal energy. Assessments of the potential for geothermal energy are increasing. A report by the National Science Foundation forecasts that the United States could be producing 137,000 Mw of electrical power from geothermal sources by 1985. In addition to electrical power, mineral extraction, fresh water production and space heating are additional potentials.

Geothermal energy potential is vast, particularly that portion which may be tapped by deep drilling technology. Numerous assessments for geothermal energy are being made. The authors have attempted to present the reader with an assessment written in general technical language. Liberal use of the literature has been made which at present is fragmented and scattered. "Geothermal Energy Technology Assessment" introduces the reader to the field and brings him up to date.

Paul N. Cheremisinoff
Angelo C. Morresi

TABLE OF CONTENTS

CHAPTER 1

WHAT IS GEOTHERMAL ENERGY

And God said, Let there be light:
and there was light. GENESIS 1:3

An interesting concept. As time passes we seem to become increasingly brash and egotistical to the extent that we begin to judge our achievements comparable to creation itself. Moving through this past century we acknowledge the swift progression of science and technology in almost all areas.

Technology has produced a standard of living for many of us which we expect and accept as our own. We fancy the ability to "create" light; to build skyscrapers thousands of feet high to reach out to the stars. Moon landings have become passe′ and space shuttles are taken for granted as had become our use and usages of energy and resources.

Economic feasibility and political indifferences have led the way for our reliance on the fossil fuels for energy. The fact that we have had no foresight for our future needs of energy and the effected oil and coal consumption rate has lead to our present dilemma. The Arab oil embargo of the winter of 1974 has created an awareness of a problem that would have become apparent within ten years, thus enabling us to get started in the search for new energy sources.

We have seen the light to some extent with respect to our environment. The National Environmental Protection Act and its amendments has effected a concern for our lives tomorrow as well as today. Thus the need for new, clean energy sources has become as necessary as oil and coal consumption result in detrimental by products to our daily lives. We note the effects of strip mining, sulfur emissions, oil spills, solid wastes and a general ravaging of the environment. The other major energy resource, nuclear energy, is not much of an improvement on the fossil fuels. Today there are multiple problems with respect to plant safety and governmental regulations; community groups, fearful of the former not being enforced and/or strict enough, disallowing construction of nuclear sites in their vicinity; and a major nemesis to nuclear power: the production of radioactive wastes that cannot be disposed of and which have half-lives of thousands of years. Furthermore, the affects of thermal pollution, as a result of the nuclear reactor, on fish, wild life and plant life as well as the climate of the earth has yet to be determined. And what few people realize is that the availability of nuclear fuel is not infinite.

With nuclear power not being the saviour of our energy needs it once was touted to be, let us examine our current and future needs of oil. United States oil fields are quite old and as every year passes the oil available from these fields is depleted until they may finally be dry wells.

For the past ten years Americans have been consuming oil at an annually increasing rate of 4.2 percent. Through 1974 our consumption rate reached 17 million barrels per day, with 37 percent of these or more than 6 million barrels per day being oil imports. The fact remains that since World War II no oil field has been discovered in the lower 48 states of the United States which contained as much as one billion barrels. Figure 1 shows the widening gap between our production and our consumption of oil over the past ten years. [1]

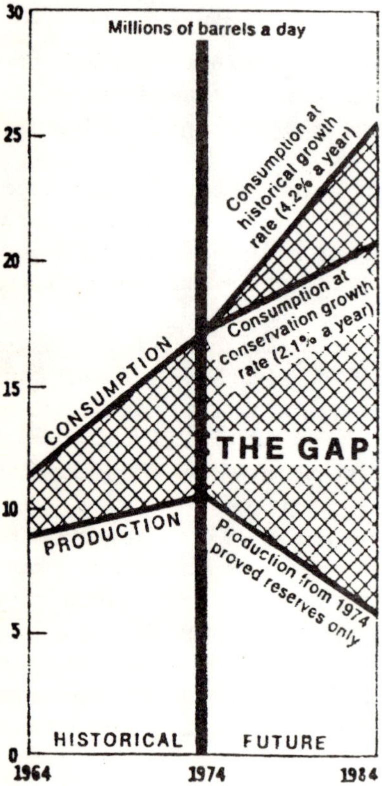

Figure 1. U.S. Oil consumption and supply.

The United States energy consumption is forecasted to continue to increase at an annual rate of 4.2 percent which each section of the economy to have a growth rate as follows: electric power development, 6.7 percent; refineries of raw materials, 5.4 percent; transportation, 3.7 percent; commercial and residential uses, 2.5 percent; and industrial uses, 2.2 percent. [2]

A conservation growth rate of 2.1 percent per year has been considered as a remedy for the energy gap, but note Figure 1 again and observe the production from the 1974 proven reserves dwindle and consumption even at the conservation growth rate effecting an alarming deficit. Conservation is not the answer, not only because we will still consume more oil than we have, but also from the standpoint of our economic development and the probable havoc to our nation caused by the stagnation of our industries.

Alternate means of energy production must be developed to enable us to proceed with the life styles to which we have become accustomed. These new energy sources must be environmentally sound to alleviate the problems associated with the fossil fuels and their effects on our society. The energy balance for the United States is shown in Figure 2. [3] The fossil fuels dominate our energy resources and if these forecasts are correct, will continue to dominate them through 1985.

The contribution of geothermal energy to our energy demand is almost negligible. Furthermore, other sources such as the sun, wind, and ocean tides do not, and

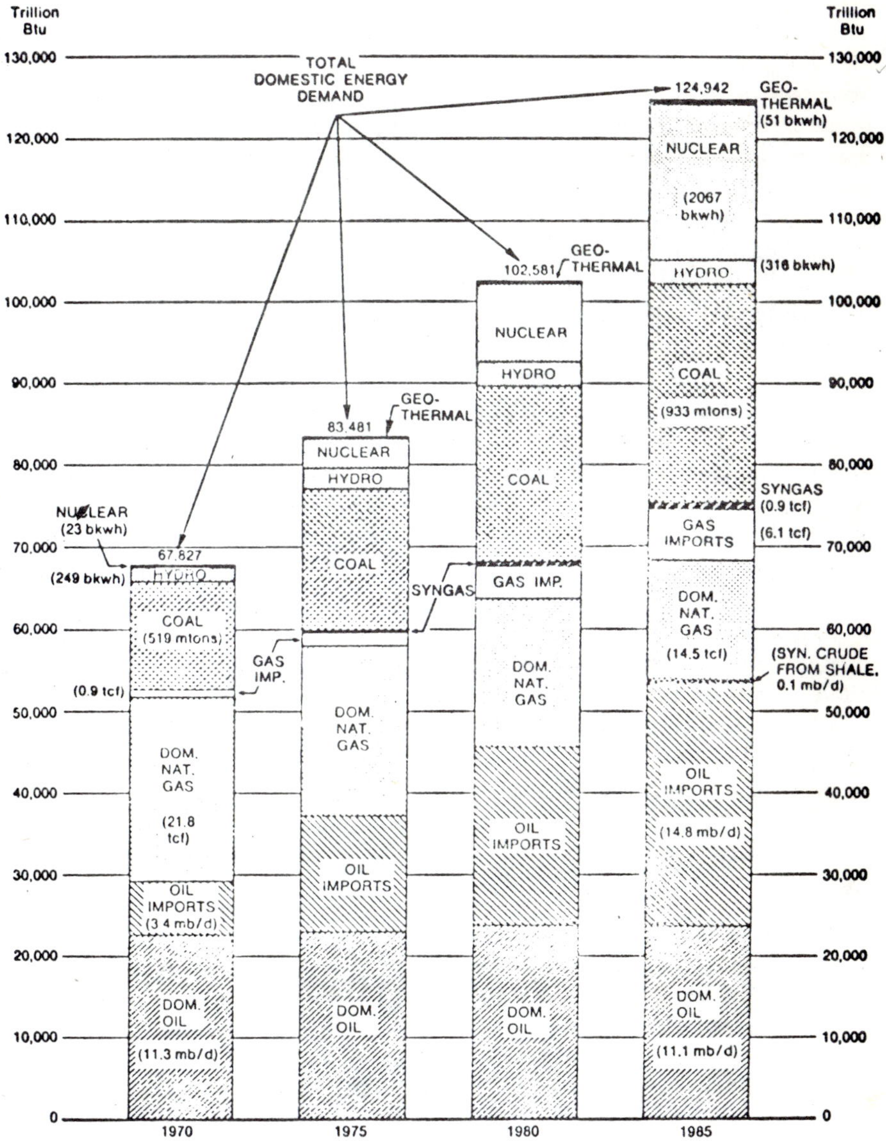

Figure 2. United States energy balance, 1972, by types and origins of resources. Categories are domestic oil (millions of barrels per day), oil imports (same units), synthetic crude oil from domestic shale (same units), domestic natural gas (trillions of cubic feet), natural gas imports (same units), syngas (natural gas produced by coal gasification, same units), domestic coal (millions of tons), hydroelectric power (billions of kilowatt-hours), nuclear power (same units), and geothermal power (same units). (Source: National Petroleum Council, "U.S. Energy Outlook: An Interim Report," November 1971.)

are not predicted to contribute accountable amounts of energy to meet our demands. Merely 3 percent of our energy needs in 1985 will be met by the resources of "new technology". These include geothermal energy, solar energy, crude oil from shale, and hydroelectric power. Nuclear power research and development is heavily financed by the governments of the world and as a result nearly 50 percent of U.S. electrical needs will be satisfied through the use of nuclear fuels. The fossil fuels, petroleum liquids, natural gas, and coal will continue to dominate the fulfilling of our power requirements.

This dependence on petroleum products does not seem realistic in lieu of the fact that proven worldwide reserves will be depleted within 70 years. [4] Nuclear energy resources are being depleted also. Therefore, monies must be allocated to explore the natural resources possibilities. They are clean, limitless and are offered to us if we are willing to research and develop the harnessing of their work.

The natural heat of the earth is geothermal energy. As a supplier of heat it is second only to another natural resource, solar energy. And like solar energy it is widely distributed over the earth's surface, but arrives in amounts too small for it to be directly converted into electricity. Available geothermal energy is concentrated in underground reservoirs, usually as steam, high temperature water, and/or hot rock.

At a depth of 6 miles the temperature of the earth is greater than 100° Celsuis, thus the total geothermal energy in storage far exceeds, by several orders of magnitude, the total thermal energy available in all the nuclear and fossil fuel resources on this plant. Solar energy is the only comparable resource of such vast quantities of energy. [5] Therefore, our energy priorities must be shifted to incorporate a resource such as geothermal energy.

History

The value of geothermal energy has been recognized for thousands of years. Waters, heated by the hot magma rock within the earth, rise to the surface and are known as hot springs. Ancient Romans and Greeks used these thermal waters as resort areas. They exploited them for their recreational and therapeutic merits. These resorts flourished throughout the Mediterranean region and extended outward to the British Isles, for the Romans utilized and developed many of the resources and processes we now take for granted. The Ancient Japanese and other peoples of the Far East also used hot springs as medicinal resorts. [6, 7]

"The gates of hell", now known as The Geysers were first acknowledged by William Bell Elliott in 1847. The Geysers are located in Sonoma County about 90 miles north of San Francisco. Mr. Elliott, while exploring and surveying the mountainous terrain of northern California, observed a valley of steam jets and discovered what is probably the largest reservoir of geothermal steam in the world. [8] Two years later the gold rush was upon the United States and this steam valley was developed into a resort with hot mineral springs. However, no attempts were made at developing this energy source until well into the twentieth century. Health spas utilizing geothermally heated waters flourished in France; Japan; Italy, and other European countries; Asia, including Russia and China; Indonesia; New Zealand; the Philipines; the Middle East; the Americas; Africa as we notice the entire world has a potentially valuable energy resource distributed not in a few oil pockets (see Chapter 3).

Not all the peoples of the world have been able to be so extravagant that they can disregard the many benefits of geothermal energy. The Italians of Tuscany have been extracting boric acid as well as other chemical and ammonium compounds from the steam jets of Larderello since the 1700's. The Larderello Co. was the first geothermal utilizing entity for it was using steam for shallow wells since 1928. Florentine Prince, Ginori-Conti ownded the company and is credited with the concept for using natural steam for the generation of power. [8]

The Larderello field first began to produce electricity in 1904 and over the next ten years the field was developed. More and more steam wells were sunk adding to the electrical output of the field. The capacity of the field has now reached nearly 400 megawatts (MW). Pioneering in the geothermal production of electricity the Italians had tried operating reciprocating engines with this steam in the 1890's and today are world leaders with respect to geothermal energy utilization.

The United States has been rather slow to exploit the resources of geothermal energy. It wasn't until the early 1920's that the United States examined the possibilities of the commercial usages of geothermal steam. Eight steam wells were drilled, but the competition from hydroelectric power was too keen to promote further development. Further complications were caused by the corrosion of piping and the steam turbines. [8] Economics are the governing factors for most enterprises and with geothermal energy they were unfeasible at the time. However, with a little hindsight we now realize a little capital investment at a loss would have saved us from many problems now.

Beppu, Japan was the site for experimental geothermal work in 1919 and these experiments effected a pilot plant in 1924 producing 1 KW of electricity. Somewhat earlier than this the Japanese began to use geo-heat [10] to heat their greenhouses.

Iceland was also an early pioneer in geo-heat. Municipal heating was effected by hot thermal waters in the 1930's and is the major source of heating today. New Zealand has been active in the development of geothermal energy around its volcanic region of North Island.

Over the past ten years many countries have begun to seriously consider geothermal energy. Research, development, construction and actual installations are now realities, although they are moving much too slow for many enthusiastic geothermalists.

Considering the United States history of exploration for geothermal energy, scientists and engineers involved in this science should be concerned with the lack of technological advancement. The problems associated with geothermal explorations have been great. Not the least of these being the fact that many discoveries are the result of explorations, geological studies, and drilling wells for the purpose of finding oil or other fossil fuels. This type of exploration is known as incidental and has led to the development of many geothermal resources – hot water wells, warm mines and tunnels; including the discovery of the Salton Sea, California geothermal brine field in 1958.

Direct geothermal exploration for geothermal resources, as noted, did begin in the United States in the early 1920's. The technology was available and did produce electricity. Later in that decade geothermal steam was sought in a field of intense fumarolic activity and mud pots. The field was located about four miles north of the Salton Sea field, but the steam yield was inadequate for power generation. As a result of these drillings, the drilling of wells to find carbon dioxide was begun in 1932. Between 1934 and 1954 the production of carbon dioxide was greater than 2.5 billion cubic feet from a total of 65 wells. On the field a dry-ice plant operated proving the commercial value of a geothermal resource in the United States.

Yellowstone National Park was the subject of research tests in 1929 and 1930. Wells were drilled in the Upper Geyser Basin and the Norris Basin and after thermodynamic and geological studies were completed, the wells were sealed. Geothermal research was undertaken at The Geysers in California, Mount Katmai, Alaska, and other fumarole areas. Wells were also drilled at Coso Hot Springs, California in an attempt to find steam in quantity and quality to produce electricity. [10] Much of the research in the early days was by private investors, thus requiring some sort of return on their investments to warrant further exploration. Most of these explorations took place in the 1920's and early 1930's. However, it wasn't until after World War II that geothermal research and development gathered any significance.

Thus, through a series of accidents, planned drillings, and geological and thermodynamic studies, the technology for the utilization of geothermal energy grew. As we proceed through the chapters in this book, we will realize the technology for geothermal resource as backward as compared to other technologies. We will try to examine the distribution and resources of the United States as well as the other countries of the geothermal world community in depth in later chapters.

Characteristics of Geothermal Energy

The distribution of thermal energy and temperatures beneath the earth's surface give rise to geothermal resources. The more widely known and accepted forms of geothermal energy, geysers, boiling pools of mud, fumaroles, hot springs, are exploited to some extent. However, a greater potential exists in regions not yet considered valuable for their energy possibilities.

Merely a generation ago the steam fields and hot mineral baths were considered freaks of nature and confined to the immediate localities of such happenings. The need for more energy has pushed exploration and the results derived thus far expose a rather extensive resource. The evidence reveals a worldwide distribution of steam and hot water reservoirs beneath the earth's surface. These energy sources have become commercially competitive in many applications with other energy forms.

As a result of these geothermal resources heat may be utilized and this heat is termed geo heat. The occurrence of geo heat is usually explained by one of the two following theories. (1) The theory that 6 billion years ago the earth was a hot molten mass of rock. This mass has been cooling through the epochs of time, with the outer crust forming as a result of a faster cooling rate. (2) The second theory presupposes that the earth is a giant furnace. The decaying of radioactive materials within the earth provides a constant heat source. Further heat is generated as the molten plastic rock masses circulate within the earth effecting frictional forces. Some combination of both theories is generally accepted.

The interior of the earth consists of a molten fluid of extremely high temperatures called magma. This magma is cooling and/or is expelling heat to the earth's surface by means of the second law of thermodynamics. The flow of heat is in the direction away from a hot sink towards a cold sink. The cold sink in this instance is the system consisting of the earth's crust, surface and atmosphere. The flow of heat outward from the magma to its dissipation in the atmosphere is a slow process. However, as evidenced from the irregularities of the earth's crust, i.e. mountains, ridges, lakes, valleys – the geological features – this dissipation of heat has produced great strains on the earth's crust. Earthquakes, volcanoes, the theory of continental drift have pushed the magma to the surface exposing geothermal regions. This hot magma contacting surface waters produces thermal waters as it cools and solidifies. Cracks caused by earth movements allow heat from the depths of the earth's interior to rise quicker and nearer to the earth's surface. These faults or cracks also allow surface waters to come in contact with the molten magma. Water flows down into the earth, is heated, and because of the high pressures existing within the earth, the water does not boil and change to steam; but its density is changed causing it to rise to the surface as a result of buoyancy forces. In addition water is contained within the molten magma and is released as the magma solidifies causing steam and/or heated water to rise to the surface. We know these surface waters or steams to be geysers, hot springs, fumaroles, or boiling mud pools. [6, 13, 14]

Figure 3 shows the geological setting of a geothermal energy source. The molten magma has been pushed near the earth's surface by the stress and strain forces from within the planet. Heat moves, by means of convection, through the crystalline rock layer to the porous rock layer. The porous rock is saturated with water which has descended through fissures in the earth's surface. The semi-permeable rock traps the heat in the rock layers below it. Being poorly conductive, rock, over the vastness of time we are discussing, can contain and isolate large quantities of heat. The water at these depths is subjected to high pressures, thus at high temperatures of maybe 500 degrees Fahrenheit, the water remains a liquid. It rises to the surface through fissures as a result of its density change and in the effect, it vents the system thereby reducing the pressure of the system. When the pressure decreases the water is allowed to boil, turn into steam and rise to the surface through fissures and/or wells. [6]

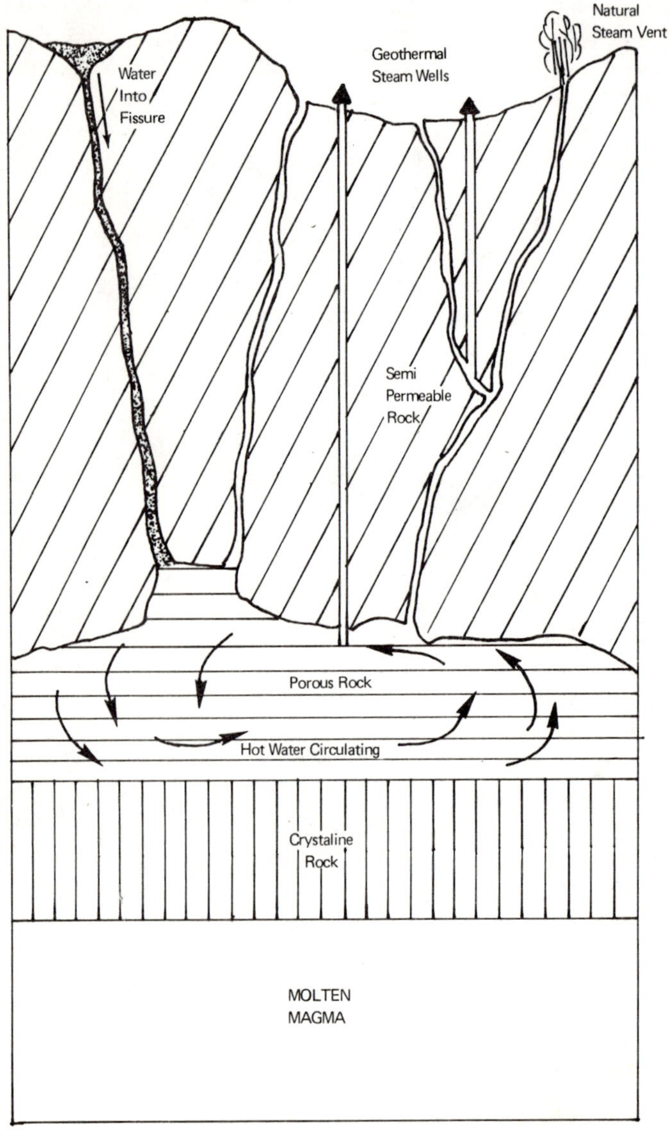

Figure 3. Geological setting of a geothermal energy source.

There are three types of resources for geothermal energy: steam, hot water, and hot rock. The steam resource can be subdivided into two types: dry steam or wet steam which is sometimes referred to as vapor dominated steam. The dry steam resource is the most desirable, in that with little pretreatment to remove corrosive materials, it can be directly piped to a turbine to produce electricity. Requirements for plant equipment are greatly reduced because there is no need for boilers, furnaces and storage tanks for fuel. The plant investment is considerably less than other electricity sources. Table 1 [15] shows the investment necessary for the four electricity sources; geothermal, nuclear, hydropower, and coal. Coal and oil plant costs are comparable since usually a fossil fuel plant is built for the eventual possibility of changing from one fuel to another.

Table 1. Cost Comparison of Four Electricity Sources.[15]

	Geothermal	Nuclear*	Hydro power	Coal*
Plant Investment, $/KW	$110.00	$225.00	$250.00	$150.00
Fixed Charges, 14%/yr/KW	15.40	31.50	35.00	21.00
Fixed Charges, mills/KWHR	1.95	4.00	6.10	4.36
Operating Costs, mills/KWHR	.25	.50	.10	.25
Energy Costs, mills/KWHR*	2.66	2.00	-	3.00
Total Costs:				
Variable Load Factor,mills/KWHR	4.86	6.50	6.20	7.61
90% Load Factor, mills/KWHR	4.86	6.50	4.55	5.92

*Costs have risen disproportionately due to fuel crisis, and safety and environmental regulations.

Orders of magnitude must be compared when discussing Table 1 since all costs are continuously on the rise and the time lapse between conception of a plant and construction completion is usually ten or more years. Note costs for coal plants have risen considerably due to EPA guidelines to rid their waste streams of SO_2 and other pollutants. Nuclear plants have been under heavy criticism for their failure to properly insure the safety of the surrounding environment, thus large monetary outlays for safety must accompany the construction costs of a plant. Therefore, a dry steam geothermal plant is the most economical by probably more than a 2 to 1 margin for the initial plant investment; and is the most economical when operating costs are compared.

The problems associated with dry steam are mainly due to plant location. Usually dry steam fields are located in isolated, highly structural geological regions thereby making plant construction extremely difficult both from a materials transportation standpoint, and from an installation viewpoint. Plants must be close to the steam source to prevent large losses in the piping and to reduce costs by requiring less piping.

*Table 2 [17]. Typical Characteristics of the Steam of a Dry Steam Field**

1.	Depth of Steam Well	> 350 m
2.	Initial Temperature	240°C
3.	Initial Pressure	35 K_g/cm^2
4.	Maximum Enthalpy	≈ 670 cal/g
5.	pH	Varies with respect to surface activity, mineral content - can range from highly acidic to slightly alkaline
6.	Mass Flow	> 1000 litres/minute

*Steam properties vary with time, with steam constituents and with well depth. Variations in the steam content of other gases produces a pressure change that is not linear. Temperatures at the well head increase with time up to a point and then start to decline. Pressures at the well decrease with time also. Enthalpies and temperatures are affected considerably by mineral content b oth quantitatively and qualitatively.

Yield from a dry steam well can reach 200,000 pounds per hour with quality characteristics of 10 atmospheres pressure at a temperature of over 400°F. [16] Because of these low pressures large amounts of steam must pass through the generating equipment. Efficiencies are about 30 percent of a conventional power plant, for in order to product 100 MW of electricity a dry steam power plant requires 455 MW of steam heat. Another parameter of plant design affected by these limited steam characteristics is size. The limiting dry steam power plant size is about 55 megawatts usually requiring the steam produced from 10 to 15 wells. The above difficulties with dry steam electricity production are a direct result of the lack of technology available. This lack of technology can be attributed to the rare occurrance of dry steam fields and the general disinterest in trying to improve this technology.

Dry steam fields of any significance are located in only five regions across the globe.

1. The Geysers, California
2. Larderello, Italy
3. Valle Caldera, New Mexico
4. Otake, Japan
5. Matsukawa, Japan

However, the relatively little exploration that is being carried out on behalf of geothermal energy may be the governing factor with regards to finding an ample amount of dry-steam fields for our energy needs. On the other hand, the above fields might be the only dry steam regions in the world and if this is the case we would do best in developing the technologies for the other geothermal resources.

Figure 4. The Geysers, California is a series of interconnecting pipelines feeding the steam from each well to the generating plant to produce electricity. Utilization of steam from a dry steam field is relatively simple and economical.

As of today explorations into geothermal resources have shown dry steam fields to be rare phenomenon and have shown wet steam fields to be much more plentiful. Wet steam fields outnumber dry steam fields by at least 20:1 and the prospects for the future are for this ratio to increase considerably.

A wet steam field is an underground reservoir of hot water under pressure. Temperatures of the water are above 100°C or the boiling point at atmospheric pressure, but the pressures at the hot water's depths prevent it from changing into steam. The field is encased in impervious rock, see Figure 3. Upon the opening of a fissure or well into the reservoir, the pressure can be released thus allowing the hot water to rise effecting a mixture of steam and water at temperatures ranging from 180 to 370 degrees Celsius. [6] This mixture is composed of 10 to 20 percent steam and hot water.

Table 3. Typical Characteristics of the Output of a Wet-Steam Field.

Depth of Well	$>$350 m
Initial Temperature	$\sim$150°C
Initial Pressure	$\sim$35 Kg/cm^2
Gas Concentration	$>$90%
Steam Concentration	Varies - relatively minor constituent of the entire well output
Other Constituents	Vary with field properties and characteristics

A wet steam well can produce quantities of steam comparable to a dry steam well, but along with the steam, water is produced which is usually 2 to 3 times as heavy as the steam. Therefore, the utilization of a wet steam field is quite different than that of a dry steam field. The operation of a wet steam plant requires centrifugal separators to separate the water from the steam before it passes on to the steam turbines to produce electricity.

Early wet steam systems used to discard the hot water into rivers or lakes. With technology in geothermalogy growing this hot water now undergoes a pressure reduction, at which point the water temperature is reduced to about 100°C as another water-steam separation is effected. These hot waters are used for heating and cooling in areas close enough to the geothermal regions to make such ventures feasible. As geothermal exploitations continues its applications become more diverse and eventually much more economical. Multipurpose projects are able to use the various components of a resource and cause the elimination of waste and a more profitable operation.

The output of a wet steam well also varies with time. The steam can condense as a result of temperature gradients in a hot water reservoir. Lower temperatures towards the outskirts of the reservoir serve to cool down the steam. As the wall is continuously tapped pressures within the system are reduced effecting temperature changes and different steam quantities.

Production from a wet steam well can generate comparable amounts of electricity as a dry steam well, however, nearly 4 times the amounts of steam and water are necessary. 400 tons of steam and water per hour at 230°C is equivalent to 100 tons of dry steam at 200°C if the steam and water is flashed (pressure is lowered to effect a separation of steam from water) at 5 atm and 154°C and at 0.8 atm and 93°C. [16] Wet steam well deposits are not as deep therefore are less expensive to drill, but produce a low pressure steam requiring more expensive equipment. Other factors must also be taken into account and these will be dealt with in later chapters.

Another type of geothermal resource is the hot-water system. Not to be confused with the hot water derived from a wet-steam system, this form is sometimes called a low-temperature field. Water temperatures are usually below the boiling point at atmospheric pressure ranging from 50°C to an upper limit of 125°C. Water from these systems is similar to the ground and surface waters of the surrounding regions and can surface as hot springs. Waters with temperatures closer to the hotter points of the spectrum surface as hot boiling springs with occasional geyser.

Surface waters and rain water enter fissures in the earth's surface and descend thousands of feet. If the subterranean rocks are highly permeable or are a network of cracks and pores, the water, heated from a hot magma source, will quickly rise to the surface and dissipate its heat. However, if the water becomes trapped between two impermeable rock layers, their heat is also contained. As is the case with all geothermal systems the rising waters through the network of cracks and pores deposit their mineral content and eventually produce a cap-rock that effects the underground hot-water reservoir. Whether self-sealing or natural these geothermal resources exhibit similar characteristics. [17] Figure 5A is a temperature versus depth curve for a typical hot-water system, Figure 5B, we note the nearness of the magma to the surface at 20,000 feet and with a temperature of nearly 1200°F at point G. Between points G and C there is a straight line relationship of temperature to depth, temperature declining with decreasing depth. At point C the water temperature is at its maximum and following it outward from the reservoirs center, we note the edges of the reservoir somewhat cooler, point B. Cooler because of the influx at this point of the cold surface waters from point A. The density changes associated with the hot water cause it to rise up a fissure in the earth's surface. There is almost an isothermal flow of water between points D and C with pressures declining. At D or about 2000 feet the pressures are low enough to allow the water to boil and from here to the surface the water temperature declines quite steadily. As mentioned previously, if the temperatures at point C or about 10000 feet are not so great, then the water will not boil at point D. Table 4 [17] shows the typical properties of the waters of a hot-water system.

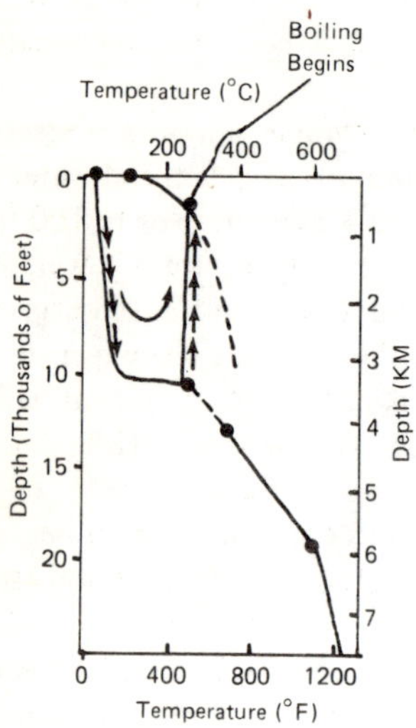

Figure 5A.

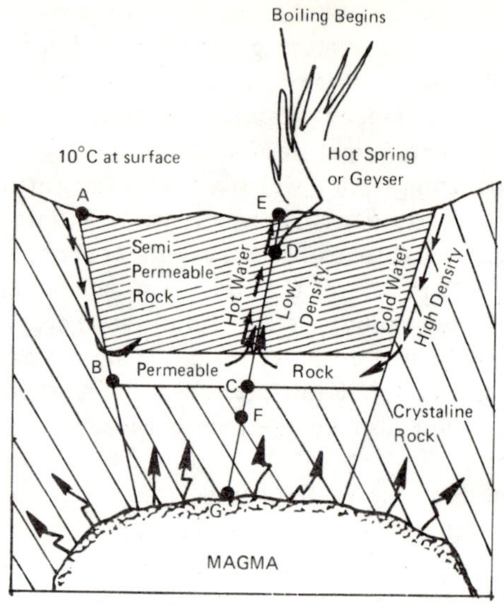

Figure 5B.

Waters from a low temperature system are most effectively used for heating and cooling. Utilization may be domestic, industrial and commercial. Heating with this geothermal resource in the USSR saved nearly 15 million tons of fuel in 1970. [6] This figure is for operations of either the experimental stage and/or small scale pilot plants. Obviously, even a small percentage utilization of just the hot water systems will effect tremendous savings of fuel.

In exploring geothermal resources the evidence has shown the world to possess numerous hot spots without the availability of natural steam and/or hot water. In fact, it is estimated that more than 50 percent [19] of the energy source of the geothermal reservoir is in hot rock of the earth and not in the fluids, water and steam. Another estimate has put the energy resources from hot rock to be more than ten times the combined total of hot water and steam resources. [18]

Hot dry rocks which are impermeable are a source of geothermal energy with heat great enough to generate electricity. The mere drilling of a well is not sufficient to gather the necessary heat in the quantities needed for electricity production. The reason: the low conductivity of rock does not allow the heat transfer to the surface to be as rapid as necessary. The answer is to effect a larger surface area of rock exposed, and to pump water into the well to "capture" this heat.

Two concepts for the above will be discussed, but both involve the creation of man-made reservoirs in the sub-surface of the planet. Artificial cavities are blown out of the earth at the bottom of the proposed well site. Means of doing this have

Table 4. Typical Characteristics of Hot-Water System.

	Description
1. Geological	Hot springs, groundwaters, underground reservoirs
2. Temperatures	50° - 125°C Waters have been recorded 150°C and because of mineral content temperatures as high as 360°C have been recorded
A. Low Temperature	Below 150°C - are not self-sealing cannot dissolve much mineral - i.e. SiO_3
B. High Temperature	Maximum temperatures 180°C - self-sealing- high in mineral content
3. Constituents	Dependent on temperature and location. Waters contents are comparable to surrounding surface waters
4. Available Form	Wells to underground reservoirs generally produce 70 to 90% mass flow water, however:

Flashing	Initial Temperature	Percent Steam
Pressure	300°C	33
	200°C	11
50 psig	150°C	0

ranged from conventional explosives to nuclear explosives. Water would then be circulated through the cavity from the surface to "gather" the heat from the hot rock. Quantities of heat gathered in this manner are questioned as to being sufficient to warrant their economic feasibility. Another problem associated with this method is the uncertainty of the underground blasts and their affect on geology of the country. Seismic waves can and do cause shifting of geological formations effecting earth tremors and other activity.

The generally accepted proposal in which there has been much interest is the idea of creating an underground reservoir by producing a crack system and circulating water through the cracks causing it to come in contact with a larger surface area of hot rock. Theories for crack creation have included explosives, both nuclear and conventional. However, in New Mexico, the Atomic Energy Commission at the Los Alamos Scientific Laboratory has been experimenting with hydrofracturing.

A bed of hard rock such as granite is cracked through the use of high pressure water. Figure 6A and Figure 6B are two schemes for the utilization of a hot rock bed. After the well or wells are drilled and the hydraulic fracturing completed, water or other heat transfer fluids is pumped down into the crack system and allowed to circulate until it has absorbed sufficient heat from the rock to be useful. The hot liquid is recovered either through the original well (Figure 6B) or through one or more recovery wells (Figure 6A). After being pumped from the underground reservoir the hot liquid would pass through surface heat exchangers and secondary fluids would run the electricity generating equipment.

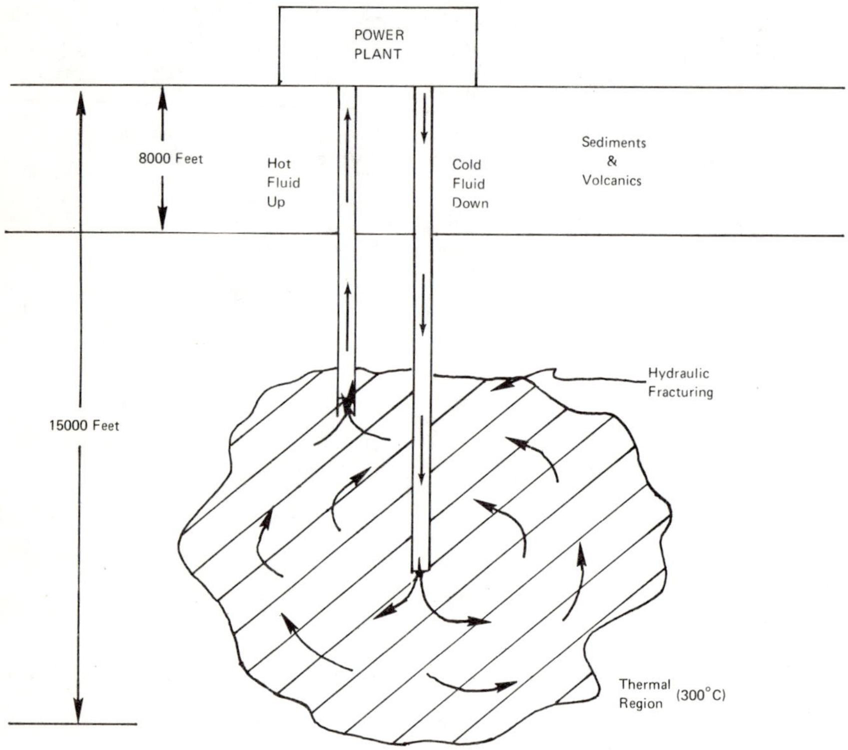

Figure 6A.

Cracks in crystalline rock formations are obtained under modest water pressures. A problem considered, but not encountered at the Los Alamos Scientific Laboratory is the water pumped down to circulate in the hydraulic cracked system would be lost to underground fissures. However, measurable amounts of water did not escape, nor did the cracks close or grow with time. Fluid pressure alone held open the cracks, or if the situation arises where holding the cracks open becomes a problem, the cracks could be propped open with particles that have been pumped down into the fissures. [6, 17, 19, 20]

Various drilling techniques have been proposed and will be discussed further in a later chapter. These proposals are for the optimum utilization of a geothermal resource. As experimentation is continued and furthered, new concepts and ideas will be examined and the exact extent and possibilities of geothermal energy could then be determined. Experiments at the LASL and at Marysville, Montana show the geothermal capacities to be economically feasible as well as abundant.

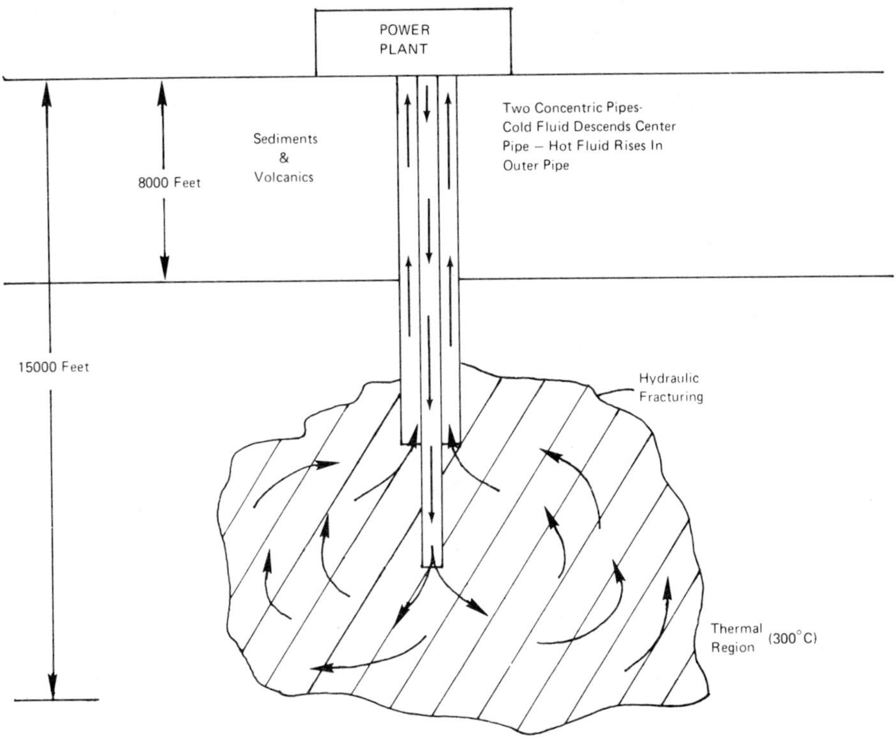

Figure 6B.

Geothermal Resources

The immenseness of the geothermal energy of the earth cannot be properly defined. Heat of the rocks of the earth above the depth of 10 KM has been estimated to be 3×10^{26} calories. However, only a fraction of this heat is available as a resource recoverable to be utilized. 0.03 percent or 10^{23} calories of the above energy total is considered hot enough and near enough to the earth's surface to be recoverable, the rest of the energy seems to be too dispersed throughout the crust.

The important thing to keep in mind is that when a number for the estimation of the geothermal resource is given, it has a certain base – whether that base be all rocks above 10 KM in the earth's crust or with a temperature above 15°C – the magnitude differences have been as high as 10^6. Obviously, the geothermal resource magnitude is dependent on several factors including physical, technological, economic, environmental, and governmental. [22, 23] The geothermal resource must be considered to be that energy available as a result of the above factors, and when one of those factors is eliminated or altered the total available geothermal resource changes by many magnitudes.

Geothermal resources can be categorized into four major classifications:

1. near-normal conductive gradients
2. abnormally high conductive gradients
3. hot-water convection systems
4. vapor-dominated convection systems

We have discussed the last three and have come to the conclusion that these sources can produce abundant quantities of energy. While not all are viable electricity producing sources, they can be used for many industrial, agricultural, commercial, and domestic needs. Where many of the so-called experts in the geothermal field have gone astray is in placing a value on the geothermal resource as to its ability to generate electricity and then determining the economic feasibility of the system.

Geothermal resources, as we shall see, can make available large quantities of fresh water, vast amounts of minerals as extracted from said waters, heating, cooling, and many uses already discussed. Thus, the whole value and availability of the total geothermal resource is incalculable and is highly dependent on man's ingenuity to utilize every aspect of this phenomonal resource.

The United States Geological Survey Research [21] explains that full utilization of the geothermal resource will be dependent on our achieving the following:

1. Advances in technology to make available low-temperature reservoirs for electrical generation.
2. Advances in drilling technology to make available geothermal reservoirs at depths greater than 3 KM, which is the present limit of geothermal drilling as economics now dictate.
3. Techniques of artificial stimulation should be developed to increase the productivity of a geothermal reservoirs.
4. Low-grade and low temperature geothermal resource use should be expanded to include space heating, product processing, agriculture, desalination, etc.

Potentialities of this resource are fantastic, availability is great, and economics are competitive or better than other sources of energy. With this in mind, we will investigate through the later chapters this geothermal resource. We believe in this first chapter we have given the reader a casual understanding of the geothermal resource and hope to develop this understanding. We further believe that a public awareness of geothermal energy and the potentialities involved with it are imperative if utilization is to occur.

CHAPTER 2

GEOTHERMAL ENERGY GEOLOGY

"And the earth was without form, and void", GENESIS 1:2

The Earth

The diameter of the earth is about 8000 miles with a circumference of 25000 miles. Its surface consists of exposed land masses known as the continents and the oceans. The surface is constantly changing as a result of erosion, earthquakes, volcanic action and other geological actions. Sediments several miles thick have accumulated, such as clays, limestones or sandstones. Other rock layers at the surface consist largely of granite and igneous rock formations. The sediments include nearly all the mineral resources we have known, i.e. oil, coal - the so-called fossil fuels.

The upper layer of rock of the earth, in whatever form it is found, is called the crust. The thickness of this crust varies across the surface of the earth with estimates ranging from 4 to 30 miles. Figure 7 shows the average crustal thickness of the crust of the Unites States. The continent's crust is composed of sedimentary rocks beneath which lies a layer of granite. The sedimentation layer thickness varies with respect to the geological time epochs. See Table 5. In North America the sedimentary layer for the Cambrian epoch and younger rocks (those rocks 600 million years old or younger) averaged only about half a mile. For all Pre-Cambrian rocks the sedimentary layer averages less than 3.5 miles. However, there are many narrow zones.[23]

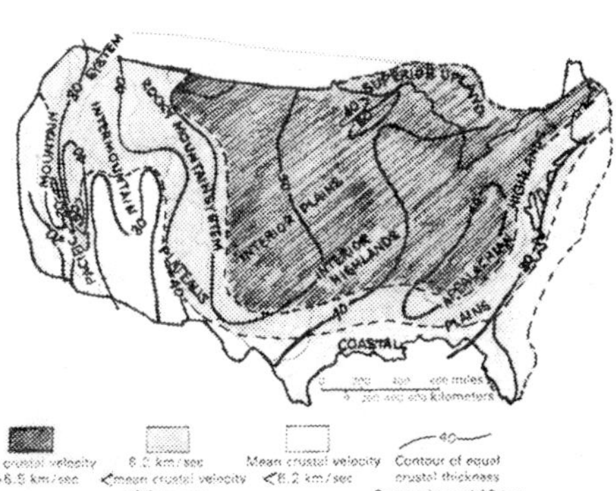

Figure 7.

Table 5. The Earth's Geological Period's.

Era	System or Period rocks time	Series or Epoch rocks time	Approx. Age mills. of years	Events
Pre-Cambrian			3500 or more	constant earth upheavals soft-bodied marine animals moderate earth upheavals
Paleozoic	Cambrian		550–600	trilobrites, mollusks first crust-aceans
	Ordovician		500	mild climate, seaweeds, clams, nautiloids, first vertebrates
	Silurian		440	mild climate, many coral reefs first life on land
	Devonian		400	varied climate, many fish amphib-ians, insects evolve
	Carboniferous		325	warm and humid, coal swamp forests, gigantic insects
	Permian		270	climate extremes, trilobites become extinct, reptiles flourish
Mesozoic	Triassic		225	arid climate, dinosaurs, mammal prototypes, first popal insects
	Jurassic		180	mild climates, giant land and sea dinosaurs, toothed birds
	Cretaceous		135	water inundates much of the earth, dinosaurs become extinct
Cenozoic	Tertiary	Paleocene	70	continent building modern plants dominance of mammals
		Eocene	60	
		Oligocene	40	
		Miocene	25	
		Pliocene	7	
	Quaternary	Pleistocene	2.0–3.0	four ice ages, many mammals become extinct, man appears
		Recent	.01	rise of civilization

The granites base of the continents were effected through a combination of metamorphism - rocks which are brought about by the changes in other rocks through heat, high pressure, and chemically active fluids - and the cooling and then freezing of molton rock. The granites consist mostly of silica and alumina. Beneath this granite layer is another rock layer composed mostly of silica and magnesium and this is called the basalt layer. Thus the crust under the continents is actually three layers of rock, see Figure 8, the sediments, the granites and the basalts.

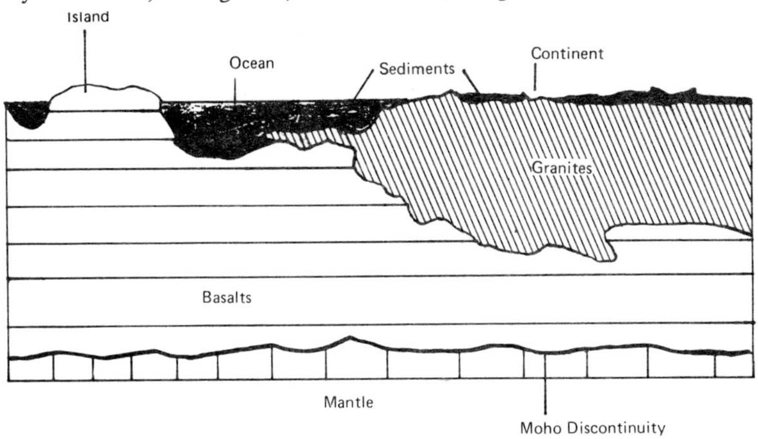

Figure 8. The earth's crust.

The crust is far from being homogeneous in terms of composition, and other properties such as elasticity and heat transfer. It is composed of much more iron, aluminum and magnesium and less silica than the surface rocks exhibit.

Table 6 shows the variance in composition between the exposed rocks and the composition of the subsurface rocks that has been estimated through the use of a seismograph and the analysis of seismic waves as they pass through the earth. Seismic waves and their resultant velocity changes as they pass through different desnities have lead to the discovery of the lower boundary of the granites and basalts. This boundary is known as the Moho discontinuity after its discoverer Andrija Mohorovicic.

This boundary encircles the earth even under the oceans where the crust is composed of a sediment layer and basalt rock layer. Basalt rock makes up the ocean islands as well as the basins. The Moho boundary lies only about 3–4 miles beneath the ocean floor, considerably less than benearth the continents. Therefore, the curst of the earth under the oceans consists of two layers, a sedimentary layer only about 75 miles thick and the basalt rock layer.

Below the Moho discontinuity is the mantle or lower layer of the earth. The mantle descends to a depth of about 1800 miles. See Figure 9. The material of the mantle is quite perplexing. It can only be examined with seismographs and the analysis of seismic wave characteristics. Depending on the time of vibration of the

Table 6. Chemical Composition of the Earth's Crust, Ocean and Atmosphere

Element	Crust Exposed	Crust Subsurface	Ocean	Atmosphere dry air to 25 Km
	%		%	%
O	46.6	43.8	85.79	21.0
Si	27.7	27.0	-	-
Al	8.1	10.0	-	-
Fe	5.0	5.9	-	-
Ca	3.6	5.1	-	-
Na	2.8	2.2	1.14	-
K	2.6	1.7	-	-
Mg	2.1	3.2	0.14	-
Ti	0.4	1.0	-	-
H	0.14		10.67	
Cl	0.03		2.07	
N	0.005		-	78.1
A	-		-	0.9
CO_2	-		-	0.03*

*Varies

mantle, it sometimes exhibits properties of a solid and other times properties of viscosity. That is, properties similar to a solid of a few hundred degrees.

The mantle's composition is generally believed to be similar to meteorites. Olivine, or iron magnesium or thosilicate, $(MgFe)_2SiO_4$, is considered the basic mantle material.

Pressures and temperatures increase within the mantle as depths increase. Viscosities and crystal structures of the mantle are changed as a result of the high pressures. The mantle becomes highly viscous at the lower depths, and at a depth of about 1800 miles a transition occurs and the liquid core is encountered.

As a result of seismic wave investigations the liquid core extends 2200 miles into the earth. Its composition is thought to be a molton alloy of nickel and iron. Within the liquid core is a solid core that is basically the same material of the liquid core but has been solidified due to the extremely high pressures. The solid core is a crystalized nickel-iron not unlike a metallic meteorite. The density of the earth increases greatly with depth to about 11.5 gms/cm^3 at the central core which has a radius of about 2200 miles.

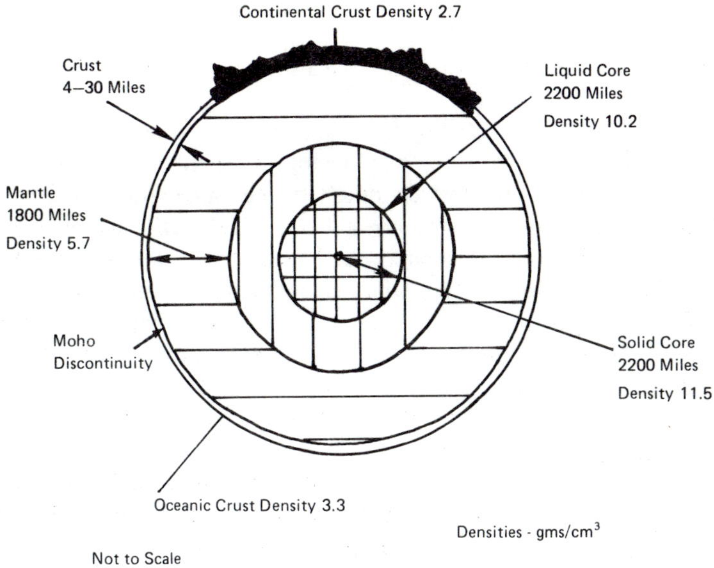

Figure 9 [25].

Very little is known about the earth's interior outside of what can be deduced through second hand investigations. However, geothermal energy investigations only scratch the earth's surface. With geothermal energy the concerns are with the earth's crust and the upper part of the mantle. Obviously the whole planet is affected by its various sectors, but even *for* the non foreseeable future, say 10,000 years, we believe for the moment we can neglect the deep interior of the earth and direct ourselves to the analysis of the earth's upper regions.

Temperature Gradient

Temperatures of the earth increase with increased depths. Across the entire earth's surface these temperature increased denote the flow of heat from the earth's interior into outerspace. (Thermodynamics: Heat flows from a higher temperature to a lower temperature body).

Two factors determine the amount of heat lost and in effect the cooling rate of the earth:

1) The temperature gradient - the rate of temperature increased with respect to depth.

2) The thermal conductivity of the rocks - the amount of heat transmitted through a one square centimeter surface in one second, under a thermal gradient of 1°C per centimeter.

Temperatures below the surface are governed by 1) the flow of heat in rocks by conduction; 2) the convective flow in circulating fluids, 3) or by mass transfer in molten magma.

When the heat flow is effected by thermal conduction, the quantities leaving the earth's surface per unit area is the product of the thermal conductivity and the temperature gradient. The thermal conductivity and temperature gradient are functions of rock properties and structures, being highly affected by topography, ground water flow, porosity, mineralogy, and fluid content of pores. See Figure 10. Note the various effects of different thermal conductivities on thermal gradients in Figure 10B. Figure 10A shows the effect of constant conductivity on the temperature gradient. Therefore, a temperature gradient at or near the surface cannot be projected downward because of the possibility of changing rock conductivity.

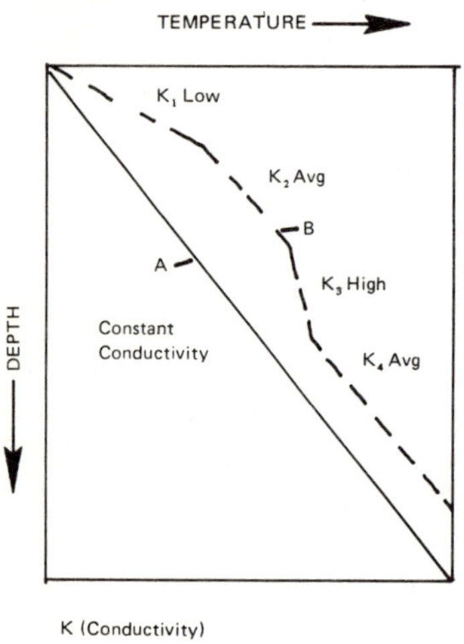

Figure 10 [17].

Temperature gradients vary across the earth's surface. The rate of increase at the hot spring and geyser area of Yellowstone National Park has been shown to reach several hundred degrees Centrigrade within a few hundred feet of the surface. In the Hungarian plains, wells have been drilled producing temperature increases of 1°C in 7 meters. However, in other areas, thousands of meters in depth, the temperature increase would cause no more than a little discomfort. The average temperature gradient of the earth is about 1°C in 100 meters. Temperature depth gradients of ten or more times normal characterize geothermal regions.

Heat flow varies with respect to time and place. Table 7 shows the thermal conductivity of most of the rocks and rock types that make up the earth's crust. The heat flow from a geothermal region would be considerably greater than a so-called normal region. This high heat flow is effected by an unusually high temperature gradient, or an unusually low thermal conductivity, or a combination of the above two factors.

The mean heat flows from the various geological regions of the earth are given in Table 8. We note the high allowances for errors involved and can conclude only that these measurements are made with unsatisfactorily inaccurate procedures. They are unsatisfactory with respect to geothermal energy resource exploration and exploitation. Here is just one area where technology has to be improved to enable scientists and engineers to better identify geothermal point sources and possible geothermal point sources.

Measurements of heat flow from the continents which are accurate and useful are few. Temperature gradients are difficult to measure also, and many factors contribute to these difficulties including the heat generated by the drilling bit and the heat contributed by hot water reservoirs which are underground and circulate,

Table 7. Thermal Conductivity of Rocks, 10^{-3} cal/cm °C [25].

Granite	6-9
Dolerite	7-8
Gneiss (normal to foliation)	5-9
Gneiss (parallel to foliation)	6-11
Quartzite	7-19
Limestone	4-7
Dolomite	9-14
Sandstone	4-11
Shale	3-6
Rock Salt	13-17
Wet Ocean Sediments	1.7-2.4

Table 8. Mean Heat Flow from Various Geological Provinces [23].

Province	Microcalories cm^2/sec
Precambrian Shields	0.92 ± 0.17
Areas of Mountains Formed During Paleozoic Time	1.23 ± 0.40
Areas of Continental Plates	1.54 ± 0.38
Areas of Mountains Formed During Post-Paleozoic Time	1.92 ± 0.49
Ocean Trends	0.99 ± 0.61
Deep Ocean Basins	1.28 ± 0.53
Midoceanic Ridges	1.82 ± 1.56
Mean for the Whole Earth	1.50 ± 0.15

taking heat from some rocks and depositing it another region. There seems to be lesser problems associated with the ocean and ocean measurement, and therefore, more accurate results are obtained from the ocean. Heat transfer effected by circulating fluids is less likely and measurements can be taken from relatively shallow depth, 2 to 5 M.

Calling for a maximum technical effort in research and the formulation of techniques suitable for the systematic measurement of thermal flow, G. Sestini[26] believes that accurate heat flow values obtained in the preliminary examinations of a geothermal field can result in the number of wells drilled being reduced when the time comes for field exploitation. When examining a geothermal field for heat flow and its ultimate utilization some of the factors considered should include the followng:

1) The relationship of actual geological structural features to heat flow and temperature gradient measurements.

2) The influence of each well site within the geothermal field.

3) The heat flow value at different depths of the wells drilled.

4) The influence of a well's diameter and filling-up system on measurements taken.

5) The approximation of obtained heat flow values by means of temperature gradients and thermal conductivity measurements.

The above stress the importance of thermal characteristics of a region and the necessity of a thorough examination of procedures before geothermal research can be advanced.

Heat Source

The following question has arisen in many a problem: Is the earth cooling or heating? We mentioned in Chapter 1 some basic theories for the heat supplied to a geothermal resource, however, let us now attempt to answer the first part of the question, namely, is the earth cooling. Was it a hot molten mass which has cooled to form the earth's crust as we now know it? There seems to be as many theories as there are questioners. The dust cloud theory is based on the belief that clouds of dust came together billions of years ago to form the earth and the various planets. The Plantisimal theory espouses the viewpoint that a large body possibly another sun came within a distance of our sun enabling it to pull materials from the sun to form the planets.

In any event, whatever the origin, the earth is believed to have had original heat. This original heat in part contributes to the average heat flow from the earth of $1.2\mu cal/cm^2 s$. In one billion years, or the time thought to cool the original molten mass to the crust as we know it, a heat flow of $1.2\ \mu\ al/cm^2 s$ would become a total heat flow of $4 \times 10^{10}\ cal/cm^2$. A heat flux of this large magnitude needs a power source of comparable magnitudes and it seems highly unlikely that the original heat of the earth could contribute a sizeable portion of this heat flux.

In his experimentation, Lord Kelvin showed that a solid earth initially at its melting point, effecting a heat flow through conduction, would have only a small fraction of the heat now observed on the earth only after 100 million years.[25] Sources of heat within the earth seem to be the answer, and everything points to radioactivity as the source.

Uranium, thorium, and potassium are the three radioactive elements which decay and give off the heat effecting the above heat flux. Some radioactivity is found to be contained in all rocks, however, these three elements and their products are not distributed in a uniform manner across the earth's surface. Uranium is in the form of two isotopes, U^{238} and u^{235}; Thorium, Th^{232}, is a single isotope; and Potasium, K^{40}, a rare isotope. Table 9 exhibits the half-lifes and heat production characteristics of the above isotopes.

Table 9. Isotope Half-Life Proportion of Heat Generation [25].

	10^9yr.	Isotope %	cal/g yr	
U^{238}	4.50	99.27	0.70	0.73
U^{235}	0.71	.72	0.03	
Th^{232}	13.9	100	0.20	
K^{40}	1.31	0.012	27×10^{-6}	

These radioactive elements are concentrated in the siliceous rocks. Uranium content in basalt averages 0.000109 percent and 0.000442 percent in siliceous granites. Potassium measurements are similar increasing from 1.07 percent in basalts to 3.70 percent in granites. Thorium quantities are usually 3 to 4 times that of uranium in a similar distribution, i.e. greater in the silicerous rocks.[23] Table 10 shows another study of these radioactive elements' concentration. It also shows the generation of heat produced in each rock type by these sources of energy.

In estimating the heat flow from the radioactive granites and other rocks for the continents and comparing it to the total heat flow from the continental masses, the mantle under the continents has been shown to supply between one-third and two-thirds of the continental heat flow. On the other hand, the mantle supplies the major portion of the oceanic heat flow because of low basaltic radioactivity. Therefore, the mantle beneath the oceans is inherently different than the mantle beneath the continents.

Table 10. Typical Heat Production in Rocks [25].

Rock Type	Concentration U (ppm)	Th (ppm)	K (%)	
Granite	4.7	20	3.4	
Basalt	0.6	2.7	0.8	
Peridotite	0.016	0.004?	0.0012	
	Heat Production (u cal/g yr)			
	U	Th	K	Total
Granite	3.4	4.0	0.9	8.3
Basalt	0.44	0.54	0.23	1.21
Peridotite	0.012	0.001	0.0003	0.013

As a result of radioactive disintegration in the crust and the mantle (the radioactivity must be concentrated in the crust for if the earth's interior were composed of the granites and other crustal rocks of radiogenic nature the earth would be a molten mass) the earth in actuality is heating up rather than cooling down.

Although the mantle is a solid there are liquid molten pockets in the upper mantle. Not much is really known about the earth's interior. With temperatures and melting points increasing with depths, it can be hypothesized that much of the mantle is at the melting temperature and the liquefied pockets are in regions of relieved pressures. Temperatures at the Mohovoricic discontinuity are estimated to be between 500°C and 700°C. Temperatures at the center of the earth are estimated to exceed 4000°C.

Geothermal hot spots are thus produced through a combination of heat sources and geological formations; and with the understanding that geothermal heat sources are not distinctly defined, we proceed to examine certain geological actions which may give us a better understanding of the workings of the geothermal resource.

Geological Actions

The flow of molten rock issued from the earth's interior is called vulcanism. Vulcanism is considered a constructive force which is building up the earth's surface crust. Volcanoes, volcanoe ridges and islands, and other volcanic rocks are products of vulcanism. Magma is hot molten rock beneath the earth's crust. Lava is hot molten rock above the earth's crust. Igneous rocks are formed by the cooling and hardening of hot molten rock in or out of the earth's surface. Sedimentary rocks are formed from the hardening of sediments derived from other rocks. Metamorphic rocks are derived from sedimentary, igneous, or other metamorphic rocks through the affects of heat, pressure, and chemically active fluids. Our purpose here is not to present an entire geology course to the reader, however, we will supply the reader with pertinent definitions from time to time.

The earth discussion we have presented thus far must be elaborated upon. Through the forces of volcanism, diastrophism (the movement of the solid part of the earth's crust), weathering, erosion, etc., the earth's surface is constantly changing. The general theory, and much evidence supports it, is that the earth's surface consists of between six and eight enormous, rigid, irregularly shaped plates, with the gaps between these plates filled by smaller plates. See Figure 11.

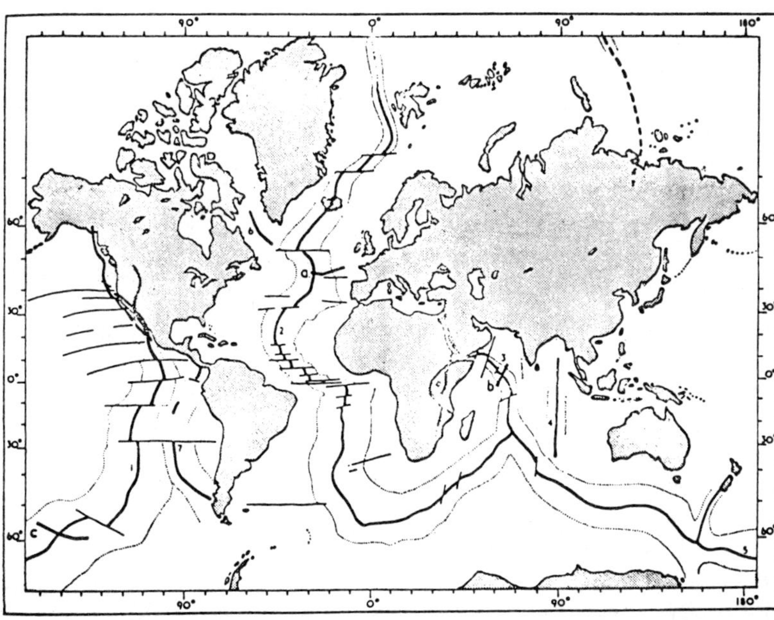

Figure 11.

Beneath these plates of the crust, as discussed previously, lies the mantle. Although the mantle is solid, it has been determined from sophisticated seismic analysis that the top 100 miles of mantle material are relatively soft and pliable. Sort of a layer of lubrication which allows for plate movement and planet development.

Marking the separation of these plates are the ocean ridges. Figure 11 shows these ridges as dark lines with traverse faults as light lines across the ridges. These ridges or submarine mountain ranges extend up and down the mid-Atlantic, around southern Africa, and up the coast of California. We note that where these ridges come in closest contact with the land masses we can recall the most recent earthquakes; i.e. Chile, California, Alaska. Earth movement during an earthquake is usually parallel to the transverse faults and perpendicular to the ridges. That is, the out pourings of lava at the ridges is the primary cause for plate separation. This is generally the case for the mid-Atlantic Oceanic ridge, for as the plates move apart on either side of this ridge, cracks are created and filled with the hot molten rock. Thus new sea floor is created as Europe and the United States move further apart each year.

This movement varies from place to place but usually averages between 1 and 10 cm/yr on each side of the ridge. Motion types also differ with respect to these ridges. Along the San Andreas fault, the Pacific plate is moving northwest relative to the Atlantic Plate and eventually will bring Los Angeles next to San Francisco in about 20 million years.[26]

What must be obvious to the reader by now is that if new sea floor is continuously being generated then the earth should be increasing in size? This is not so. As the Atlantic Ocean becomes larger, the Pacific Ocean is in effect shrinking as California moves closer to Japan. The sea floor of the Pacific Ocean is being forced into the earth. Figure 12 tries best to show the action of the forces on the plates and their resultant movement. The pulling of the Pacific plate into the earth's interior is being effected around the entire Pacific Ocean. From Alaska and the Aleutian islands to Japan, up and down the coast of the Americas, the Mariana, the Phillipines, and the southwest islands of the Pacific including New Zealand and Indonesia. Lavas are usually present in these areas and it has been suggested that this lava is formed as a result of the sinking plates melting.

Let us review Figure 11 again and we see the direct relationship between these ridge areas and geothermal areas. In Chapter 3 we will discuss the distribution of the geothermal resource in more detail. However, let us examine the mid-Atlantic ridge. It cuts right through Iceland and on Iceland there is much volcanic activity and many hot springs and geysers. The ridge passes through the southwest Pacific south of Australia and New Zealand where many hot springs are known to occur. In the northern Pacific Japan and the Aleutians have reported much geothermal activity. Geothermal resources have been developed in western United States and explored up and down the Andes. All these geothermal areas of the Pacific are thought to be associated with the sinking Pacific plate. Another ridge passes through southern South America and geothermal resources have been identified in the area.

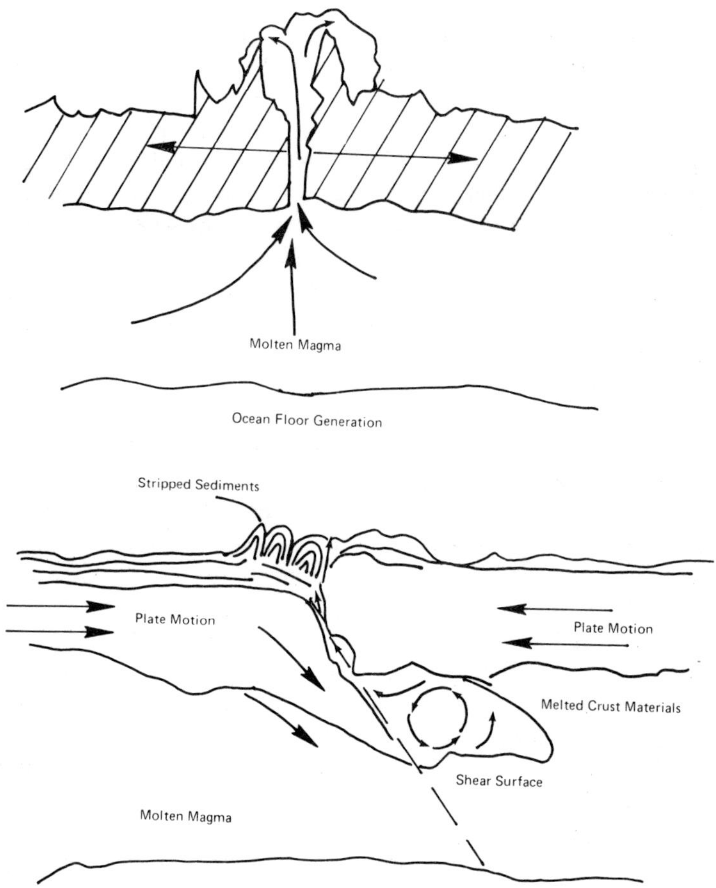

Figure 12. Ocean plate sinking.

Plates are colliding with one another all over the world. Africa is pushing up into southern Europe producing the Alps and probably the geothermal areas of Italy, Hungary and the rest of Europe's hot spots. India is pushing into southern Asia effecting the Himalayas and Siberia into Russia producing the Urals. In all these areas we note plate movement and the availability of the geothermal resource. The above hypothesis can also explain the Rocky Mountains, and California and Nevada energy sources.

To further suggest this relationship of geothermal sources with plate movement, investigate the regions where geothermal resources have not been identified. The east coast of the United States, the West coast of Europe and the east and west coasts of Africa are not subjected to certain geological actions because in these regions the continents and ocean floors are part of the same plate. In these regions there are no volcanoes, no earthquakes (which are clearly associated with plate interaction), no plate shifting, and no geothermal resources[25] as we now know them.

Geological activity such as vulcanism and earthquakes are the result of diastrophism. One of the many effects of this diastrophism is the geothermal regions of the world. Whether it be where an outpouring of lava generates new mountains and ocean floors or the destruction of a plate wherold material is being melting to form new lavas coming to the surface, these regions exhibit geothermal properties. Therefore, exploration for a geothermal source of energy must be carried out by means of a scientific approach and investigating geological activities with sophisticated instruments rather than looking for surface evidence such as a geyser or hot spring.

Gilluly, Waters, and Woodford[23] explain the continental drift theory in detail and note many theories which attempt to justify continental drift. We will summarize one of the more generally accepted theories: Conduction and convection are the two methods that a fluid heated from below loses heat. Heat flow through rock as a result of thermal conduction is rather slow, thus as the heat in the molten rock is contained due to a low heat loss rate, and increased due to continuous heating from below a temperature gradient develops effecting a fluid motion of the liquid rock. This motion is favored in a low viscosity. The hot molten rock rises as the cooler denser rock sinks back to the heat source. The hot molten rock is then cooled and sinks. Thermal convection is thus effected. Low viscosity, low conductivity, and large volumes are the favorable parameters for convection.

The mantle is this hot liquid rock; its motion is quite slow, its viscosity high and it acts as a liquid if it is examined over a period of thousands of years. Therefore, the mantle is a liquid base on which the crust floats and flows. Figure 13 is a scheme of the convective system of the earth. The hot molten liquid rises pushing materials out through the mid-Atlantic ridge and in the east Pacific (NOTE the formation of the Hawaiian Islands) and sinks under the shear surface which runs along the Pacific coast of the Americas

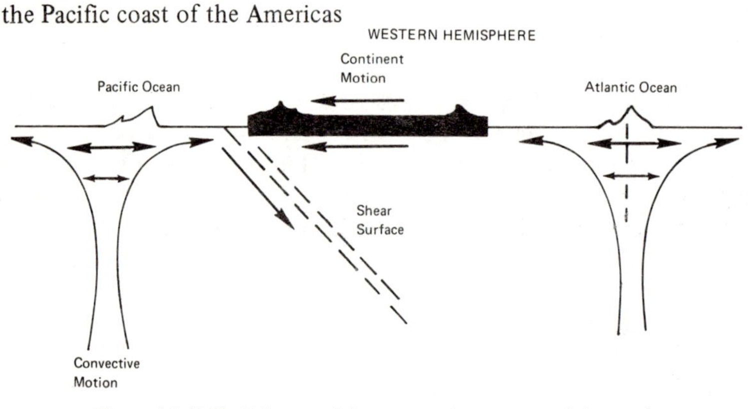

Figure 13 [23]. Scheme of the convective system of the earth.

Thus mantle convective motion is the mechansim of bringing the geothermal energy from its source to the surface. It is also the mechanism for ocean floor and island generation as well as plate sinking and continent motion.

Bullard[25] further explains the mechanism for heat transfer to the surface of the earth. Upon rising to the surface of the earth many processes occur within the molten rock and they include:

1) pressure drop
2) temperature drop
3) liquids and gases come out of solution
4) crystallization begins

The lava and the igneous rocks it produces are dependent on the above four parameters. That is the solid substances evolving from the freezing lava will vary with respect to freezing point and whether that freezing point has been reached rapidly, such as the process of quenching, or quite slowly. Igneous rocks also vary with pressure differences and material content of the lava such as water and other volotiles. The crystallization of the lava begins first with the formation of quartz. The first crystals will then sink to the bottom of the molten material and more heat will be given off as more crystals begin to form. Hot water and steam and other minerals are released from the solution of molten rock.

Heat is then released at the surface as these hot gases and hot waters and steam find their way up fissures. Many suggestions and theories have been made concerning the heat transfer from the interior to the exterior of the earth. We've noted conduction through the crust, convection of the mantle and then conduction to the surface or the spewing of mantle material on the surface or the melting of crust material as a result of plate movement. Heat is transferred also by convection in a subterranean water reservoir and by the releasing of these waters on the surface.

The origins of these waters and therefore the origin of the most basic geothermal heat exchangers are disputed. Some aruge that the waters are derived from the molten material and as described above are released upon cooling. Others dispute this theory on the basis that the quantities of water involved are too great for such an hypothesis. These people seem to lean more towards our description in Chapter 1 (See Figure 3) that surface waters trickle down into pervious rock, are heated, and then rise to the surface.[28] The most realistic assumption is that thermal waters and steam originate through a combination of the above theories but with surface waters accounting for a significantly larger proportion of the thermal waters. [23,28] (Figure 14).

We now see geothermal energy, not as a strange phenomena nor as William Ben Elliott put it when he discovered The Geysers of California, "The Gates of Hell", but as the resultant of patterned geological actions. Geothermal regions are not located randomly around the world but are a natural resource that have been effected by geological processes. Not unlike the fossil fuels they are distributed across the world in those regions where the forces of nature have accounted for them. Earthquakes, vulcanism and continental drift are as much a part of a geothermal resource as a geyser or hot spring. In fact the geyser and hot springs can be likened to the top of an iceberg. They are just the end result of a phenomenal amount of parameters interrelating to produce a geothermal resource.

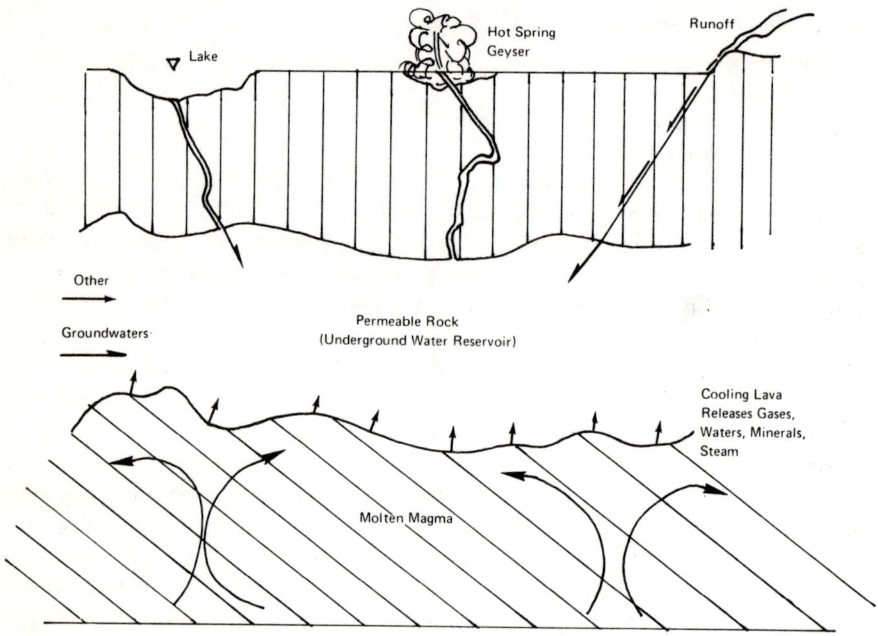

Figure 14.

CHAPTER 3

GEOTHERMAL ENERGY DISTRIBUTION AND AVAILABILITY

He causeth the vapours to ascend from the ends of the earth . . . PSALM 135:7

Heat concentration zones have developed across the globe as a result of continental drift and related geo-actions. We can describe these zones as geothermal heat belts. These heat belts are where the geothermal resources of the world are available for exploitation. The six types of geothermal resources include:

1) dry steam fields
2) wet steam fields
3) low temperature fields
4) hot rock areas
5) geopressure fields or geopressure zones
6) areas in and around volcanoes

All the above have been discussed in previous chapters with the last two types of resources being variations of the first four.

In considering the availability of a geothermal resource the type of energy source and its corresponding geothermal field and geological formations must be regarded as basic factors. Also to be considered are the geothermal constituents. Each geothermal resource, while the same from the viewpoint that each produces heat are different in many respects. The makeup of a geothermal resource can include one or more of the following in various proportions; steam; hotwater; gases in solution with the steam and water; sinter and precipitates in geothermal areas; mineral content of the geothermal host rocks; heat in the geoformation; natural gas energy in the form of compressed water. And finally, the actual ongoing geothermal processes which generate the natural heat and the water recharge must be investigated.

The United States

It has been estimated that a 110 MW geothermal steam plant would eliminate 10^6 barrels of foreign oil imports. As a result of this and the growing worldwide pressures to find alternate energy sources, Congress passed the Geothermal Energy Research Development and Demonstration Act of 1974. Its goal is to assess the nature and size of the geothermal resource base. Previous to this act Congress passed the Geothermal Steam Act of 1970 which greatly increased the development of geothermal resources in western United States. Known Geothermal Resources Areas (KGRA) on public domain in the western United States have been estimated to be nearly 1.5×10^6 acres or 6×10^9 square meters.[30] Most of these regions are in the vicinity of the geological structures and activity that was identified as basic to a geothermal resource in the previous chapter.

Figure 15 shows the western thirteen states and springs and fumaroles with temperatures above 160°F. There are at least 100 of these hyperthermal regions in these states; regions where the surface temperatures are high suggesting a boiling fluid beneath the surface. These areas are concentrated in the geothermal belt of eastern California, central and western Nevada, Oregon, southern Idaho, northwestern Wyoming and Montana. Another belt lies across Utah and still another runs through Colorado and New Mexico. The Aleutian Islands of Alaska have much hyperthermal activity as does Hawaii.

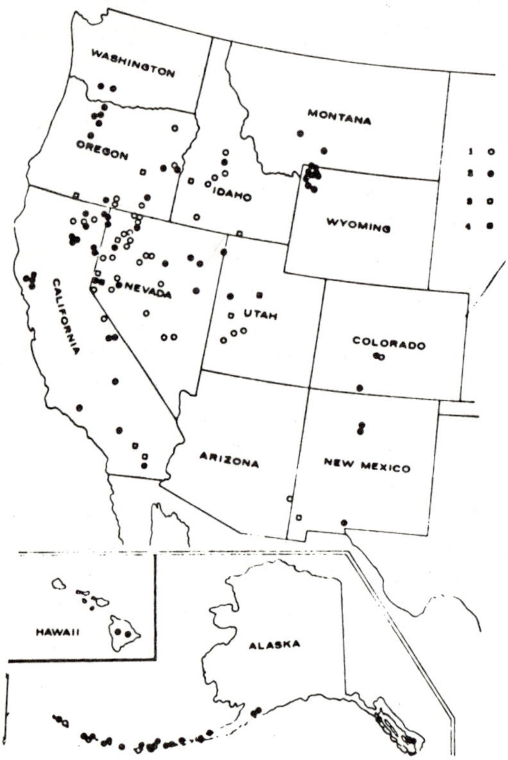

Nomenclature

- ○ Single spring or fumarole
- ● Multiple springs and fumaroles
- □ Other basins of hot ground water
- ■ Mine shafts

Figure 15. Known hyperthermal regions of the Western United States.

Of the geothermal resources in the United States, over 90 percent are located in these western thirteen states. Along with the 100 hyperthermal regions there are approximately 1000 other geothermal localities of warm springs and fumaroles which range in temperature from just 20°F above the average annual air temperature to the hyperthermal temperatures that may exceed the boiling point at a particular location.

Geothermal power generation in the United States has only been available since 1960 when the Pacific Gas & Electric Company put its first commercial unit on line at The Geysers, California. This electricity is generated from geothermal steam. Because of technical and legal problems other producable geothermal areas have not been developed or are just now being investigated. These areas include Casa Diablo, Salton Sea, California; Beowawe, Brady's Hot Springs, Nevada; and Yellowstone National Park, Wyoming. Most of these areas have base temperatures of between 350°F and 400°F. They, like the geothermal drillings listed in Table 11 are hot fluid or steam resources. Hot rock geothermal resource investigation and development are under way in Marysville, Montana where the temperatures reach 700–800°F. at 6000 feet. Therefore, while much of our geothermal energy energies are expended towards research of hot fluid sources as seen from Table 11, there is more hope and more availability of the ultimate utilization of the hot rock or dry geothermal energy source.

The Geysers, California

In 1960 the steam from eleven wells was combined producing a 12,500 KW generating unit. Successful implementation of these wells has resulted in the rapid expansion of the Geysers field for electricity production. Cost estimates and engineering studies proved the feasibility of using small geothermal steam turbine-generator units. New facilities are continuously being constructed, as The Geysers have now surpassed Larderello, Italy as the largest electricity generating geothermal field. Known data has been analyzed and estimates of The Geysers geothermal field have been for ultimate production to reach 1000 kW.

Wells of varying depths have been drilled. The first wells were drilled to depths of between 200 ft. to 1000 ft. and steam flows ranged from 40,000 to 80,000 lb/hr. As geo-technology improved, higher pressure and deeper wells were drilled. Wells over the last 10 years vary from 2000 ft. to 7000 ft. in depth. 9000 ft. wells have been drilled and produced as much as 380,000 lb/hr of steam. More typical steam flow rates are in the 200,000 lb/hr range.

Figure 16 shows a map of the Geysers area. The various units that are and will be in production are spread out over a 7 mile zone. The Geysers are a geological formation of fractured gray wastes, shales and basalts formed during the Jurassic and Cretaceous ages. A large reservoir exists as a result of permeability due to fault, shear and fracture zones. Temperatures of reservoir are of the 500–550°F and shut-in pressures of wells deeper than 2000 feet are between 450–480 psig. The

Table 11. Geothermal Drillings for Steam & Hot Fluids to October 1969".

	Field or prospect	Number of wells drilled	Dates drilled	Purpose	Operator	Surface phenomena, °F	Drilling data: Maximum T, °F	Drilling data: Maximum depth, ft	Miscellaneous
1.	**The Geysers, California**	78	1921-25 1955-69	G	Geyser Development Co.; Magma Power Co.; Thermal Power Co.; Union Oil Co.; Geothermal Resources International; Signal Oil and Gas Co.	several areas of fumaroles and hot springs; to 212°	$>500°$	8,175	Enthalpy $\sim$1200 BTU/lb; steam phase in reservoir; superheat in shallow wells; P.G. & E. generates 83 MW of electric power; 110 MW under construction; 7-mile zone proven productive by drilling; field capacity may exceed 1,000 MW.
2.	**Salton Sea (Niland), California**	15 ($>$65)	1927 (1932-54) 1957-65	G (CO_2) O, G	Pioneer Development Co. (3; 1927); Kent Imperial Oil Co. (1; 1957); O'Neill Geothermal Co., Inc. (2; 1961-62); Transcontinental Power Co. (1; 1961); Shell Oil Co. (2; 1963-64); Western Geothermal, Inc. (2; 1962-64); Earth Energy, Inc. (3; 1963-64); Imperial Thermal Products (1; 1965).	boiling mud pots; CO_2 seeps; to 212°	680°	8,100	Enthalpy $\sim$450 BTU/lb; brine contains 220,000-260,000 ppm. chlorides of Na, K, Ca; $\sim$20% flashover to steam; problems of corrosion, scaling, waste disposal; 3 MW pilot generator; pilot recovery of salts; within structural rift zone; other geothermal pools believed present to south; extensive shallow CO_2 field, produced 1932-54.
3.	**Casa Diablo (Mammoth), California**	11	1959-69	G	Magma Power Co. (and associates).	several areas of fumaroles and hot springs; to 194°	352°	1,063	Hot water with 5% steam flashover; reservoir T$>$370° F; problems of waste disposal and scaling; at edge of extensive structural depression.
4.	**Brady's Hot Springs, Nevada**	9	1959-65	G	Magma Power Co. (and associates) (9); Earth Energy, Inc. (2).	boiling springs and steam seeps; to 209°	418°	5,060	Hot water with 5% steam flashower; reservoir T$>$400° F; problem of scaling; at edge of major structural depression.
5.	**Beowawe, Nevada**	11	1959-65	G	Magma Power Co. (and associates).	fumaroles, steaming ground, hot and geysering springs; to 205°	414°	2,052	Hot water with 5-10% steam flashover; reservoir T$>$400° F; problems of inflow of cold water and scaling.
6.	**Yellowstone National Park, Wyoming**	15	1929-30 1967-69	R	Carnegie Institution of Washington (2); U.S. Geological Survey (13).	very extensive areas of geysers, boiling springs, fumaroles, steaming ground; to 205°	$>460°$	1,088	Dry steam and hot water with steam flashover; main reservoir T$>$460° F; probably no development possible within National Park; 6 areas explored along 40 mile zone; major caldera.
7.	**Steamboat Springs, Nevada**	36	1920-62	G, R, W	Magma Power Co. (and associates) ($\geq$8; 1959-62); U.S. Geological Survey (8; 1950-51); many private parties.	boiling and geysering springs and fumaroles; to 205°	369°	1,830	Hot water with 5-10% steam flashover; reservoir T$\sim$350° F; severe problem of scaling; hot water found in about 120 wells along 12-mile zone north to Reno by shallow drilling.
8.	Clear Lake (Sulphur Bank Mine), California	4	1961-64	G	Magma Power Co. (and associates); Earth Energy, Inc.	hot springs, to 120°; hot mine waters, to 176° at 310 feet	367°	$\sim$5,000	Hot water with 5% steam flashover; shallow CO_2 horizons; reservoir T$\sim$360° F; severe problem of waste disposal.
9.	Valles Caldera (Jemez Springs; Sulphur Springs), New Mexico	4	1960-64	O, G	Westates Petroleum Co. (1); Dunigan Tool and Supply Co. (3).	hot springs, to 187°	$>$400°	5,600	Hot water with 5-10% steam flashover; production reported to decrease with time; reservoir T$\sim$400°; at edge of major caldera.
10.	Surprise Valley (Lake City), California	4	1959-62	G	Magma Power Co. (and associates).	mud pots and ebulating springs; mud eruption in 1951; to 208°	320°	2,150	Hot water with steam flashover; reservoir T$>$300° F; at edge of major graben.
11.	Calistoga, California	3	1960-61	G	Calistoga Power Co.	hot springs, to 173°	278°	$>$2,000	Hot water with minor steam flashover; reservoir T$\sim$280° F; problems of disposal and scaling; very many shallow hot water wells along 2-mile zone one geysering well, erupting to 40 feet.

	Field or prospect	Number of wells drilled	Dates drilled	Purpose	Operator	Surface phenomena, °F	Drilling data: Maximum T, °F	Drilling data: Maximum depth, ft	Miscellaneous
12.	Darrough, Nevada	1	1962, 1963 redrill	G	Magma Power Co. (and associates).	boiling springs, to 207°	265°	812	Very large flow of hot water, with minor steam.
13.	Wilbur Springs, California	2	1965, 1968	G	Magma Power Co; Cordero Mining Co., and Worldwide Geothermal Exploration Co.	hot springs, to 153°	285°	3,717	Very large flow of hot water, with minor steam; 30,000 ppm T.D.S.
14.	Mount Lassen (Terminal Geyser), California	1	1962	G	Geysers Steam Co.	geyser and hot springs, to 204°	264°	1,285	At edge of National Park; extensive fumarole and hot springs areas, within Park, associated with caldera.
15.	Arrowhead Springs, California	2	1963	G	R. A. Rowan and Co.	hot springs, to 187°	234°	571	Hot water
16.	Wabuska, Nevada	3	1959	G	Magma Power Co. (and associates).	hot springs, to 162°	222°	2,223	Hot water
17.	Monte Neva, Nevada	1	1965	G	Magma Power Co. (and associates).	hot springs, to 193°	190°	402	Hot water
18.	Stillwater, Nevada	1	1964	G	O' Neill Geothermal, Inc., and Charles Oliphant Co.	none	~240°	>3,200	Large flow of hot water: 270 foot deep well flows at 190 °F; within Carson Sink structural depression.
19.	Cedarville, California	1	1962	G	Magma Power Co. (and associates).	hot springs, to 130°	130°	734	Hot water
20.	Wendel, California	1	1962	G	Magma Power Co. (and associates).	hot springs, to 208°	174°	630	Hot water
21.	Amedee, California	3	1962	G	Magma Power Co. (and associates).	hot springs, to 208°	225°	1,116	Hot water
22.	Kelley's Hot Spring, California	1	1969	G	Geothermal Resources International.	hot springs, to 204°	230°	3,206	Major lost circulation zone at 1,600 feet.
23.	Crump (Adel), Oregon	1	1959	G	Magma Power Co. (and associates).	hot springs, to 164°	~250°	1,684	Geysering well, yielding 400 to 600 gal/min; eruptions reaching 100 to 150 feet.
24.	Lakeview, Oregon	2	1961	G, W	Magma Power Co. (and associates).	hot springs, to 185°	~240°	~ 650	Geysering wells, eruptions reaching 50 to 60 feet. Use of water in greenhouses.
25.	Puna (Kalapana, Pahoa), Hawaii	5	1961	G	Magma Power Co. (and associates).	steam seeps; hot ground; to 182°	218°	692	Steam seeps, associated with 1959-60 eruptions of Kilauea, in eastern rift zone.
26.	Fales Hot Spring, California	1	1962	G	Magma Power Co. (and associates).	hot springs, to 141°	<100°	413	Warm water
27.	Bridgeport, California	1	1962	G	Magma Power Co. (and associates).	hot springs, to 171°	123°	982	Hot water
28.	Tecopa, California	1	1962	G	Magma Power Co. (and associates).	hot springs, to 109°	low	422	Warm water
29.	Coso Hot Springs, California	1	1967	R	U. S. Navy	fumaroles, boiling springs, shallow steam wells; to 208°	287°	375	Well can be induced into geysering; no sustained flow; Holocene perlite domes and basalt flows.
30.	Randsburg, California	1	1960	G	Magma Power Co. (and associates).	none	240°	772	1920 borehole for mercury encountered steam.
31.	Fort Bragg, California	2	1965	G	Northern Plains Petroleum Co.	none	148°	3,977	Site located by geochemical prospecting and air photo interpretation.

	Field or prospect	Number of wells drilled	Dates drilled	Purpose	Operator	Surface phenomena, °F	Drilling data: Maximum T, °F	Drilling data: Maximum depth, ft.	Miscellaneous
32.	Genoa, Nevada	2	1962	G	U.S. Steel Corporation.	hot springs, to 160°	181°	1,268	Approximately 20 shallow, temperature-gradient holes drilled in area.
33.	Smith Valley, Nevada	1?	1962?	G	U.S. Steel Corporation.	hot springs, to 143°	—	—	Well(s) reported to have been drilled at Simpson Hot Springs.
34.	Pyramid Lake, Nevada	3	1964	G	Western Geothermal, Inc.	boiling springs, to 208°	~240°	5,888	Large flow of hot water.
35.	Fernley (Hazen), Nevada	1	1962	G	Magma Power Co. (and associates).	hot springs, to 204°	270°	750	Area of steaming ground 12 miles to east; at edge of Carson Sink structural depression.
36.	Crescent Valley, Nevada	1	1965	G	Magma Power Co. (and associates).	hot springs, to 122°	166°	410	Hot water.
37.	Klamath Falls, Oregon	> 350	c. 1920-1969	WS	Klamath Falls School District; very many private parties.	hot springs, to 185°	>235°	1,805	One low-pressure dry steam well; hot water used to heat over 400 buildings, via heat exchange system: reservoir T~230° at about 400 feet; original hot springs gone, via lowering of water table.
38.	Ward Ranch (Fly Ranch), Nevada	1	c. 1920	W	———	warm and hot springs	>220°	shallow	Continuously erupting well, has built 15 foot high travertine cone; at edge of Black Rock Desert structural depression.
39.	Burns, Oregon	1	——	W	———	warm springs, to 80°	~200°	~ 700	Water used to heat greenhouse.
	Temperature data for selected areas of hot ground water								
	Desert Hot Springs, California	~ 80	1934-68	W	Desert Hot Springs Water District; and very many private parties.	none	~200°	807	7-mile zone of hot water wells, along trace of San Andreas fault; reservoir T~200° at ~200 feet.
	Animas Valley, New Mexico	3	1954 and earlier	W	———	none	216°	106	One well can be induced into geysering; other warm and hot water wells in general region.
	Raft River Basin, Idaho	4	1935-48	W	———	none	211°	540	Flowing wells, also called Frazier Hot Spring.
	Boise, Idaho	2	1890?	WS	Boise Water Corporation.	hot springs, to 140°	176°	~ 400	Water used for space heating in over 200 homes.
	Temperature data for selected mining districts								
	Comstock Lode (Nevada), silver and gold mining district					temperatures to 178 °F at 3,080 feet			eight miles SE of Steamboat Springs area
	East Tintic (Utah), copper, silver, lead and gold mining district					temperatures to 163 °F in drill holes about 1,500 feet deep; may reach 200 °F to SE			

Symbols

Purpose of well

G geothermal exploration
W water for domestic or irrigation use
R research into geothermal systems
O oil or gas test
CO_2 exploration for carbon dioxide
S water for space heating

reservoir is thought to be composed of steam and water and total capacity of The Geyser area cannot be determined. Many more drillings have been initiated in the outskirts of the field and thus the reservoir may be even more extensive than previously anticipated.

Imperial Valley, California

The Imperial Valley lies in the Gulf of California and just over the East Pacific Rise and the North American Continent. Sediments saturated with water are nearly 20,000 feet thick with a heat flow to the surface greater than the worldwide average. Some heat flux regions of the Valley produce a heat flow over 10 times the average. The porosity and permeability of the rocks and deposits of this geological formation are effected by the extremely high temperatures at the depths of the basin and as a result dissolved solids content has reached 260,000 mg/1.

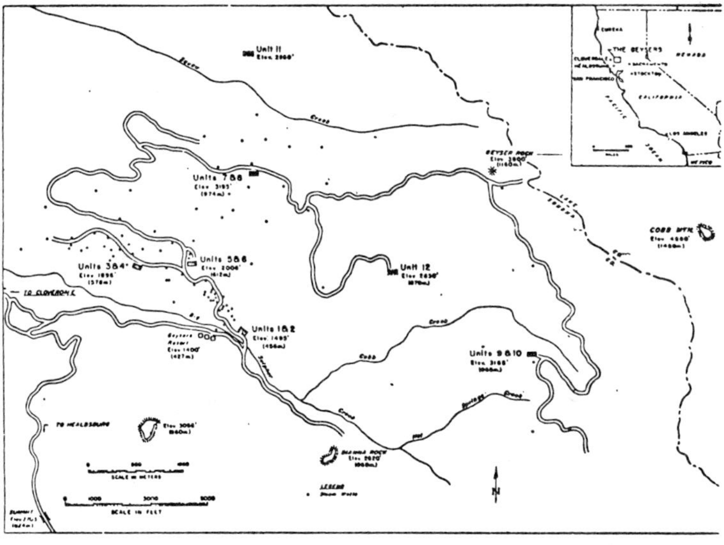

Figure 16. Map of The Geysers, California [31].

Recoverable groundwater, negelcting cost estimates and with a dissolved solid content of less than 35,000 mg/1, is estimated at 1.1 billion acre-ft. with 20 percent of this water available at a temperature of 150°C or greater. Dissolved solids content must be kept at a minimum, since they increase the pretreatment water costs tremendously. Usuable recoverable water from the field is limited to 100 million acre-ft. because of high energy costs and low well yields. Almost 62 percent of the recoverable water originates from the Colorado River and the rest from local sources.[22]

The exploration program of the Imperial Valley consists of a combined system for electricity production, water desalination for agriculture use. The system will be entirely closed with depleted heat fluids being pumped back into the reservoir hot rocks for secondary use. The problems that have arisen with this and similar systems are environmental (what to do with dissolved solid residues) and will be discussed in a later chapter.

Yellowstone National Park

The largest and hottest geothermal region in the United States is probably located at Yellowstone. The whole geothermal district extends for 40 miles, but is subdivided into lesser fields by natural effects. In the north is the Mammoth Hot Springs. Lying 15 miles to the south is the Norris Geyser Basin which contains about a dozen geysers and many hyperthermal pools. Further south are zones of fumaroles, hot springs, geysers, and other geothermal activity. Included in this display of geological fury is the familiar Old Faithful Geyser.

Thirteen wells have been sunk at Yellowstone to a relatively shallow depth (i.e. just over 1000 ft.) and base temperatures are estimated at 500°F to 520°F. Dry steam has been effected from one well and others have produced hot water which has flashed to steam. Ash flow formations of 1,500,000 years are found at Yellowstone National Park along with basalt flows, obsidian, glacial debris, diatomaceous sediments suggesting a volcanic period which extended into the Holocene time. These, however, comes a point where man has no right to exploit and destory an occurance which is not his and whose beauty has been in existence since before he walked the earth. Many people agree with this philosophy and therefore it seems highly improbably that the National Park will be commercially developed for geothermal utilization.

Other Hot Fluid Regions

Other potentially viable systems of geothermal resources have been discovered at Steamboat Springs, Nevada; Clear Lake, California; Vailles Caldera, New Mexico; and Surprise Valley, California. Figure 16 corresponds to Table 11 and shows many of the geothermal investigations to date. Like many of those resources previously mentioned, the fluids obtained from wells drilled in these areas varies in quantity, quality and temperatures. At one well at Steamboat Springs temperatures reached 340°F at 350 feet. Problems have arisen from carbon dioxide in the vapor state causing the incrustation of calcium carbonate within the wells and a severe reduction in steam flow. Clear Lake reports temperatures in excess of 365°F at 4000 feet producing 50,000 pounds of steam per hour and 1.5 million pounds of water per hour at 100 psig. A high boron content in the water creates an environmental disposal problem. Most of the fluid sites have comparable problems and the obvious answer is technology. Water treatment techniques of the geothermal waters must be developed to encompass the high temperature; high dissolved solids, mineral and chemical contents of these waters and to provide for safe disposal methods. However, what must be repeated - the utilization of each part of the geothermal resource will bring costs for the whole project down to the point where exploitation of each entity, (steam, hot water, minerals, fresh water, etc.), can be profitably processed.

Hot fluid regions are found in central and eastern Alaska, Idaho and northern Nevada which preliminary studies show to have extremely high potentials as geothermal energy sources. Geothermal fluids are being used in Oregon for space heating. 350 wells were drilled at Klamath Falls to supply heat to buildings. The fluids pass through a heat exchanger with minicipal waters which flow through buildings. Other current uses of geothermal heat in Oregon include greenhouses, resorts, baths, farm buildings and schools. Thus, it must be realized that even waters of relatively low temperatures - 185°F - can and should be utilized.

Geothermal resources are not available only in the form of hot fluids as most of the early investigations suggest.

Dry hot rock and other forms of the geothermal resource are available and are being investigated.

Dry Geothermal Reserves

One of the major problems associated with this type of heat analomy is to classify it and find it. Exploration techniques are rather crude, that is, they are more or less examinations of surface activity as evidences to what lies below. Areas which may have geothermal energy readily available go unclassified as the surface is literally just scratched.

Twenty miles northwest of Helena, Montana is an old mining town, Marysville. Temperature measurments in old mine shafts and wells discovered what is now the highest temperature gradient in the United States. The temperature at Marysville, rises 10°F for every 1000 feet. At this gradient temperatures will reach 700–800°F at 6000 ft. All this is happening in an area where there are not hot springs or geothermal surface evidence within 20 miles.[32]

The site is about 12 square miles in area and the geology of the region shows no indications of hot water availability. The source of heat is believed to be a tear shaped mass of hot, dry, impermeable rock. Methods for harnessing the heat from a dry rock heat source have and will be discussed. 2.5 million dollars has been allocated for Marysville research, [33] but the Marysville project is the infancy of the resource. From Figure 16 it can be seen that the major portion of the known geothermal districts in the United States warrant further exploration. Regions sim-

Figure 16A. Geothermal Resources of the United States. Well numbers correspond to wells sited in Table 11 [11].

ilar to Marysville probably exist in many parts of the world and in those very regions of the United States that have not been examined in detail. Research into dry geothermal depcsits is being carried out at the Atomic Energy Commission's Loss Alamos Scientific Laboratory and they hope to reach temperatures of 300°C at depths of over 7000 feet. Thus, the availability of dry hot rock is apparent and improved exploration techniques will change this availability to utilization.

Salt Domes, a Geothermal Resource [34]

Salt domes have been noted throughout the world for their proximity to oil and gas deposites and for the mining of salt and potash. The most interesting development concerning salt domes is their eventual value as a source of geothermal heat. Salt domes are a geological heat anomaly, and this is better seen in Table 12 where the thermal conductivity of rock salt is shown to be nearly twice as high as most other rocks. Therefore, the salt dome can bring heat up from deep within the earth.

Table 12. General Comparison of Rock Types to Rock Salt.

Rock Type	Thermal Conductivity X 10^{-3} cal/cms°C
Sedimentary rocks	1.0–8.0
Metamorphic rocks	5.2–8.4
Igneous rocks	3.1–9.8
Rock Salt	17

A typical salt dome is the Lovann formation of the Gulf Coast. In Figure 17 the salt dome of Lovann formation origin is shown. Temperatures of 460°F have been recorded at a depth of 14,552 feet and estimated at 570°F at 25,000 feet. These depths are far greater than any previously available geothermal resource, therefore, economic feasibility must be considered before this source can be thoroughly explored.

Temperature gradients vary to about .5°C/meter or 2.6°F/100 ft producing a heat flow that ranged between 6.2×10^{-6} cal/cm^2-sec to 10.1×10^{-6} cal/cm^2-sec. The average heat flow of the surrounding area is 1.2×10^{-6} cal/cm^2-sec or nearly 6 to 10 times less than the salt dome anomaly. Heat flow through a geological structure is calculated in the following manner:

Heat Flow = thermal conductivity (K) N temperature gradient

or

$$\frac{\text{calories}}{\text{area} \times \text{time}} = \frac{\text{calories}}{\text{distance} \times \text{time} \times \text{temperature change}} \times \frac{\text{temperature change}}{\text{distance}}$$

The heat from a salt dome is available in a number of ways and they will be discussed in our systems section.

The salt domes prove there are many sources of geothermal energy and these will be available to us if we widen the scope of our investigations to include the heretofore impossible and impractical.

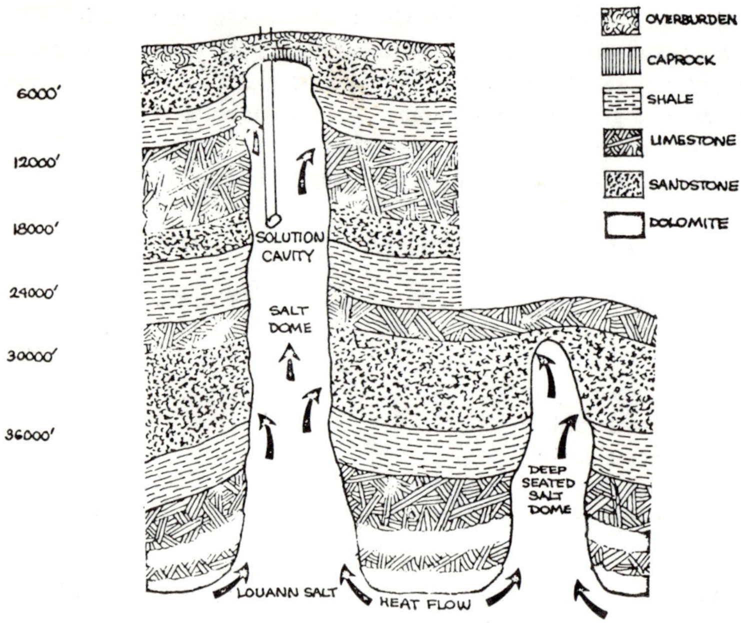

Figure 17. Salt dome as a geothermal heat source.

The World

Figure 18 is a map of the world showing the regions of high geothermal resource potential. The distribution coincides generally with the ridge phenomena examined in the previous chapter. Many of the western industrialized nations, which have become dependent on foreign fuel exports have been looking at geothermal energy and have decided to examine its many possibilities. However, even countries like the U.S.S.R. which has huge oil reserves, have begun to investigate geothermal resources. In fact, the Soviet Union announced plans to build the world's largest geothermal electricity generating plant.

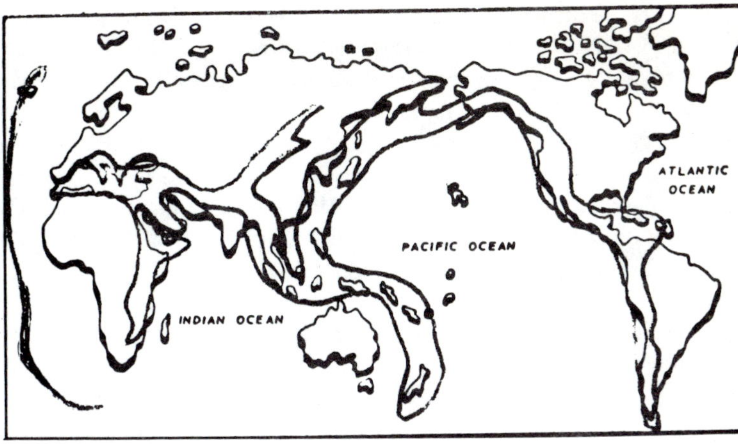

Figure 18. Geothermal regions of the world coincide generally with the geothermal plate movement and activity.

The world status of geothermal exploration and development is rapidly changing. Table 13 notes those countries which have expressed a geothermal interest and the various stages their investigations have taken them.

Table 13. Extent of Geothermal Exploration and Development, 1972[9].

Nation	Electric-power generation/ construction	Experimental power stations	Significant direct utilization	Other geothermal-field discoveries	Additional exploration under way[c]
Chile				x	
China		x			
Ethiopia					x
Guadeloupe (Fr. W. Indies)				x	
Hungary			x		
Iceland	x		x	x	
Indonesia					x
Italy	x				x
Japan	x	x[b]	x	x	x
Kenya			x		x
Mexico	x[a]	x	x	x	x
New Zealand	x		x	x	
Nicaragua				x	
Philippines		x		x	
El Salvador	x[a]			x	
Taiwan				x	
Turkey				x	
U.S.S.R.	x	x	x	x	x
United States	x	x[b]	x	x	x
Zaire		x[b]			

[a] Under construction. [b] Inactive.
[c] Other geothermal exploration/interest: Algeria, Argentina, Bulgaria, Burundi, Colombia, Costa Rica, Czechoslovakia, Ecuador, Fiji Islands, Greece, Guatemala, India, Israel, Malawi, Mali, Morocco, New Britain, New Hebrides, Peru, Poland, Portugal (Azores Is.), Rwanda, Spain (Canary Is.), Tanzania, Tunisia, TFAI (French Somaliland), Uganda, Venezuela, Yugoslavia, Zambia.

Italy

Since the first geothermal generating plant went into service in 1913 in the Lardello field in Italy, the Italians have pioneered geothermal utilization and technology. The capacity of the generating complex has grown from the original 250 kW to over 390,000 kW. The famous Lardello field has 13 plants producing 365,000 kW and another 25,000 kW capacity is generated at Monte Amiata, 75 miles south (region 5 in Figure 5).

Research and exploitation at the Lardello-Monte Amiata is continuous. Over 500 wells have been drilled at this complex to an average depth of 500 m. The average well yield is 23,000 kg/hr at Lardello and 36,000 at Mone Amiata. This is consistent with the fact that both mass flow and pressures decline with time and that Monte Amiata has been in production for a much shorter time period. However, evidence of declining pressures and yields has led geologists to believe the field has been tapped to capacity. Still, research continues through various surface prospects and surveys with geological, geochemical and geophysical means throughout the entire region and surrounding areas. The entire region is also being explored by other methods including infrared imaging to determine if any geothermal energy sources have been overlooked.

Geothermal exploration is being carried out in other regions of Italy for possibility of finding hot fluids for steam production. These regions include Monte Volsini, Monti Aimini, Monti Sabatini, and Colli Albani. Areas of northern Italy have been explored and rejected, but southern areas in the vicinity of Mount Vesuvius (Compania ovest) have been investigated. Fluids from these areas are considered hot but are not dry steam generators. Therefore, alternate means of geothermal utilization including alternate methods of electricity production will have to be developed.

Progress in Italy in the field of geothermal energy has to be attributed to technology improvement and the discovery of new energy sources. Technology improvement in both areas of exploration and discovery and in steam production through the use of more efficient turbines. Early turbines were of the noncondensable type which consume twice as much steam per Kwhr as do the condensing steam turbines, 20 kg/Kwhr vs 10 kg/Kwhr. Thus the more effective utilization of the steam can increase a field's electricity output without any increase in steam output.

Probably the most important reason for Italy's progress in the geothermal area is the committment by Italy in form of the following policy towards geothermal energy development. [35]

1. The maintenance of the primary energy sources in the new operating steam fields. The natural decrease in flow rate of wells which have been contributing steam for a considerable number of years will be compensated for by the introduction of new wells.
2. The exploitation of the surrounding areas of the known steam fields to

100

Figure 19. Geothermal research areas of Italy.

increase the overall steam production.

3. The identification of new sources of geothermal energy through researchers developed at the national level.
4. The widespread use of small and simple turbo-alternator exhausting to atmosphere generating units which are to be installed at the site of the wells drilled in peripheral regions so the new steam can be immediately utilized.
5. The development of engineering technology to obtain the most efficient operating pressure.
6. Increase output and reduce steam input into generating stations through

improvement of plant and auxilliary equipment.

7. Improvements of steam pipeline networks including insulation systems.
8. The decentralization of power stations. This objective serves a twofold purpose: (a) to reduce plant operating costs with the installation of remote control systems, and (b) to reduce pipeline costs and therefore thermodynamic losses.

Iceland

Iceland, unlike Italy which imports nearly 80 percent of its oil, has an overabundance of electricity from its hydroelectric installations. However, Iceland has been using thermal spring waters for centuries and has significantly tapped their geothermal resources for a variety of purposes. Throughout this small island nation, located on the mid-Atlantic ridge, geothermal waters heat schools, social centers, greenhouses and homes. The waters are also used domestically in both private and public buildings. Other uses include drying and refining diatomaceous earth and there has been expressed a desire to use it in heavy water production, freeze drying fish products, and large-scale flower cultivation for export.

Over 100,000 people of Iceland receive geothermal heat for their homes and 90 percent of the homes in the capital city of Reykjavik use geothermal waters. Contributing to this widespread useage is that (1) the water requires very little processing before it is distributed, and (2) the hot water is carried for 10 miles in concrete encased steel pipes with a loss in temperature of only 9°F (5°C). Mineral content is extremely low for geothermal waters. 400 ppm of dissolved solids are found, of which 35 percent are silicon oxide which is commonly found in drinking water. Natural fluoride of the proper proportions for drinking water and nitrogen are also contained in the waters. Dissolved sulfur compounds while harmless, are noticeable, but considered an acceptable inconvenience. The water is passed through gas separators as it leaves the ground to remove dissolved gases.

Iceland does have a 3000 kW geothermal power station in northern Iceland at Namafjall. The use of this power and other geothermal energies is equivalent to burning 180,000 metric tons of fuel oil or 1170 million KW-hr of electrical power. Coupled with the financial benefits are the environmental benefits. It is estimated that burning fuel oil for one year would produce over 2000 tons of sulfur dioxide, 850 tons of nitrogen oxide and 210 tons of flyash.

New investigations are being conducted throughout Iceland as new methods of exploitation as well as new areas for exploration are developed to increase the use of this natural resource.

Mexico

Mexico's Federal Electricity Commission convinced by much geothermal surface activity, carried out an intense geological, geophysical, and geochemical investigation of the area surrounding Cerro Prieto. Figure 20 locates Cerro Prieto in a region that is being heavily investigated and exploited on the northern side of the border by the United States. It is part of a thermal field which has been effected by the San Andreas fault network. Reservoir temperatures are in excess of 300°C and the depth of the average well is 4000 ft.

Figure 20. The San Andreas fault system extends into Mexico forming a geothermal field which stretches between the U.S. and Mexico.[37]

22 wells have been drilled to produce a high temperature hot water system that supplies flash-steam powered electric generating plants. The field is being expanded and hopes are for it to reach a capacity of 370,000 kW by 1976. Figure 21 and its corresponding Table 14 show the availability of the geothermal resources in Mexico. Los Negritos, Ixlan de Los Hervores, Los Humeros, Los Azufres, La Prinavera are located on a major fault system called Chapal-Acamby. These areas offer fantastic potentials for Mexico, but their utilization of this major natural resource has been too slow and actually a dereliction of responsibility has occurred as only within the past ten years has the Chapala-Acamby fault system been explored for possible exploitation.

Japan [38, 39]

It is estimated Japan will be importing nearly 90 percent of her energy resources by 1980. Therefore, Japan has turned inward to utilize all her natural resources including geothermal energy. In 1966 Japan constructed a 20,000 KW geothermal power plant in northeast Japan in Matsukawa. Another power plant was built in the

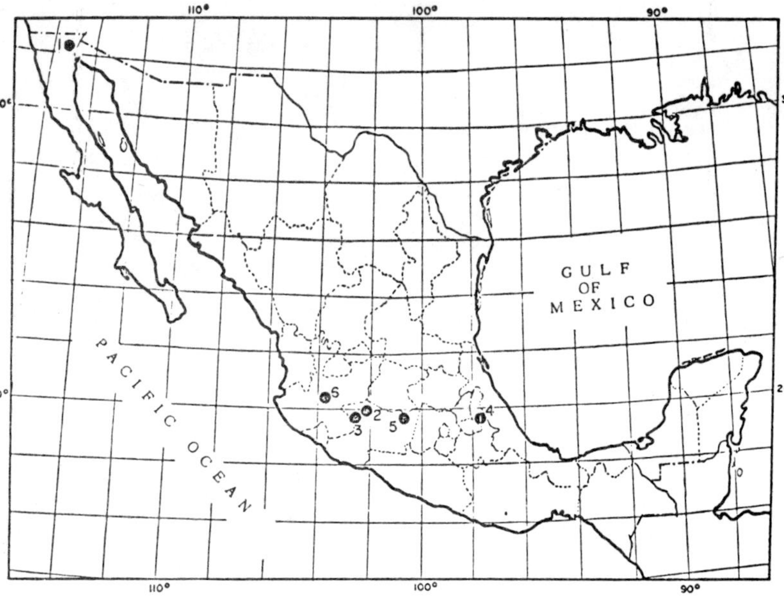

Figure 21. Main geothermal fields.[36]

Table 14. Characteristics of the Geothermal Fields[36].

Name of Field	*Physiographic Expression*	*Type of Activity*	*Surface Temperatures*
CERRO PRIETO (88 m a.m.s.l.) Field under development for 75 MW	Delta valley, with surrounding farmlands.	Fumaroles, hot springs, mud volcanoes, lakes, and salt deposits.	45-100°C, with high values in the wells
LOS NEGRITOS (1533 m a.m.s.l.)	Caldera-like structural valley, open towards NW. Flows of Tertiary and Quaternary lavas, surrounded by farmlands.	Mud volcanoes and bubbling lakes aligned NE-SW	36-90°C
IXTLAN DE LOS HERVORES (1535 m a.m.s.l.)	E-W faulted structural valley, with flows of basalt and andesite. Acid soils unfit for cultivation in the hydrotermal area.	Hot springs, boiling mud lakes, steam discharges, and low-pressure wells. Deposits of geyserite with ceramic residues.	45-100°C
LOS HUMEROS (1800 m a.m.s.l.)	Centrally faulted caldera. Explosion craters. Deposits of basalts, sands, and stratified tuffs.	Low-pressure fumaroles, hot ground, and alteration zones	85-90°C
LOS AZUFRES (2800 m a.m.s.l.)	Intensely fractured double caldera. Altered rhyolites and volcanic ash.	Superheated steam (fumaroles), large mud volcanoes, hot springs, and bubbling lakes.	45-110°C
LA PRIMAVERA (1570 m a.m.s.l.)	Faulted caldera. The geothermal area is located on two N-S faults which bound a graben.	Superheated steam (fumaroles), alteration zones and hot ground.	55-100°C

Otake area (see Figure 22) with a capacity of 16,000 kW in 1967. Japan established in 1960 the Japan Geothermal Energy Association to promote geothermal investigations. Its results lead to the selection of the ten areas shown in Figure 22. Hot springs and fumaroles are plentiful throughout Japan as volcanism is highly active. Contributing to a large geothermal reserve is an annual precipitation rate of 160 cm or 600 tons of water. Thus, Japan could have a geothermal reserve of up to 10 million kW as technology improves.

Key

1. Kutchoso
2. North Hachimanta
3. Matsukawa
4. Takinokami
5. Onikobe
6. Nasu
7. Oshirakawa
8. Otake
9. Takenoyu
10. Ibusuki

Figure 22. Geothermal exploration areas of Japan.

Hot springs are being used for medicinal purposes and recreational facilities. Agriculture and animal breeding also benefit from thermal waters. However, exploitation of these waters for heating purposes have been not nearly as extensive as in Iceland. Many of Japan's problems relating to the utilization of its geothermal resources are social rather than technical. Since their earliest beginnings the Japanese have been using these thermal waters for bathing. Hot springs are fully developed throughout Japan as resort areas and others have been made into national parks. Thus, geothermal exploitation will be further slowed as new technology tries to displace or at least coexist with ancient traditions.

U.S.S.R.

The U.S.S.R. is one of the few major countries of the world to make a major committment to geothermal energy. The Soviet Union's program includes scientific investigations of theoretical and applied geothermics. The present work includes:

1. The study of deep thermal processes.
2. The examination of regional distributions and the condition of formations of the geothermal field in regions where direct measurements can be taken.
3. Technology improvements and development with respect to both apparatus and exploration.
4. The practical use of deep heat.

Geothermal resources vary across the Asian Continent and the condition of these resources vary from region to region within the U.S.S.R. Table 15 shows the distribution of thermal waters in the different regions of the U.S.S.R. Some of the volcanic regions have rocks and waters with temperatures of 100°C at the surface. At a depth of 100–400 meters some regions of Kamchatka and the Kuril Islands have temperatures of 150°–200°C. It is known that between 50–60 percent of U.S.S.R. territory contains economically exploitable thermal waters. These geothermal resources are comparable to the combined oil, gas, coal and peat resources of the U.S.S.R. The geothermal water reserve is estimated at 7.9×10^6 m^3 per twenty-four hours at a depth of 1000–3500 meters and with temperatures at 50–130°C. And 70 percent of this reserve is 1500 m from the surface.

Table 15. Thermal Water Distribution Across the U.S.S.R.[40]

Region	Thermal water reserves m^3/day × 10^3	Part of the thermal water reserves %
1 - European part of the USSR among them:	1210	15.4
Caucasus	780	
Crimea (with E Carpathian region)	330	
2 - Central Asia	550	7.0
3 - Kazakstan	480	6.1
4 - West Siberia	4300	54.3
5 - East Siberia and the Far East	1360	17.2
among them:		
the south of east Siberia	300	
the northeast (Jakut, Magadan region, the Chukotsk coast)		
·Khabarovsk region	360	
Sakhalin	100	
Kamchatka and the Kuril Islands	600	

What is more important than availability is application, and the U.S.S.R. has investigated many possibilities for its thermal resources:

1. Heating – including domestic and industrial use of hot water and cold water supplies.
2. Agriculture – cultivation of vegetables through hotbed and hothouse heating; and for the needs of cattle breeding.
3. Chemical – chemical separation from thermal waters for use in a variety of areas including balneological (theuraputic effects from thermal waters) purposes.
4. Electricity – energy production.

New Zealand

Current exploitation of the geothermal resource by New Zealand encompasses three regions. Wairakei, where a 160,000 kW geothermal power station is located; Kawerau, where a paper and pulp mill uses geothermal steam for processing and electricity production; and Rotorua which uses geothermal energy extensively for heating purposes.

Electric power production is expected by 1976 from a new field currently under investigation, the Broadlands. Table 16 gives some characteristics of the Broadlands geothermal field as well as other selected fields of the world. The Broadlands locality and the other geothermal regions of New Zealand are situated on Figure 23.

Table 16. Selected Geothermal Fields of the World and Their Characteristics.[9]

Field	Reservoir temperature, °C	Reservoir fluid	Enthalpy, cal/g	Average well depth, meters	Fluid salinity, ppm	Mass flow per well, kg/hr	Non-condensable gases, %
Larderello	245	Steam	690	1,000	<1,000	23,000	5
The Geysers	245	Steam	670	2,500	<1,000	70,000	1
Matsukawa	230	Mostly steam	550	1,100	<1,000	50,000	<1
Otake	200+	Water	~400	500	~4,000	100,000	<1
Wairakei	270	Water	280	1,000	12,000	—	<1
Broadlands	280	Water	400+	1,300	—	150,000	~6
Pauzhetsk	200	Water	195	600	3,000	60,000	—
Cerro Prieto	300+	Water	265	1,500	~15,000	230,000	~1
Niland	300+	Brine	240	1,300	260,000	~200,000	<1
Ahuachapán	230	Water	235	1,000	10,000	320,000	~1
Hveragerdi	260	Water	220	800	~1,000	250,000	~1
Reykjanes	280	Brine	275	1,750	~40,000	~400,000	~1
Namafjall	280	Water	260	900	~4,000	400,000	6

Nearly all of New Zealand's geothermal development has been in the volcanic area of Taupo in the North Island. Waters high in sodium chloride content with temperatures of 270°C. High mineral content of the waters has presented many problems including corrosion, and calcite deposits reducing outputs.

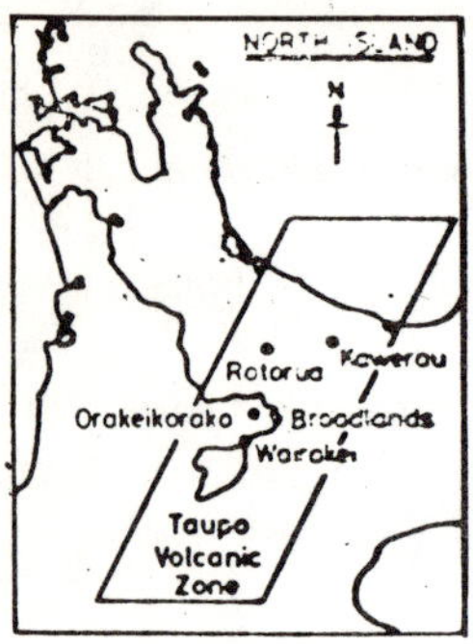

Figure 23. The North Island geothermal area of New Zealand.[42]

Other World Resources

To discuss the individual workings of each installation of each nation's geothermal program is beyond the scope of this paper. We have tried to acknowledge the major accomplishments and resources of some of the globe's members and to give the reader a feel for the abundance of this valuable natural resource and what leadership with direction and foresight can contribute to our needs.

Geothermal resources of Ethiopia [43], Indonesia [49], Western Anatolia [45], Nicaragua [46], Taiwan [47], Columbia [48], Guatemala [49], Costa Rica [50], Uganda [51], Tanzania [52], India [53], Czechoslovakia [54], Yugoslavia [55], Hungary [56], Poland [57], and those cited in Table 13 are developed to various degrees. Some are in preliminary study stages while others are generating electricity and developing more and more applications for this resource. Figure 24 shows the geothermal electric power stations of the world in their various stages of development. Figure 25 shows regions of the world where geothermal energy has a significant use other than for the generation of electricity.

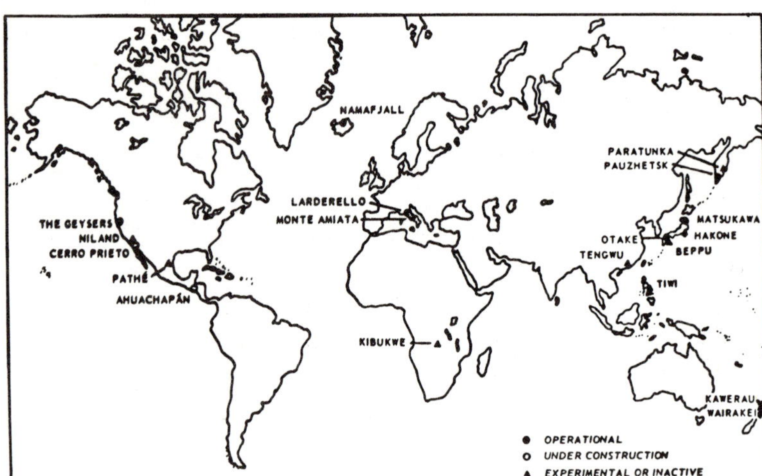

Figure 24. Regions of the world where geothermal energy generates electricity.

However, the current energy crises has produced such an energy strain on various nations that alternate sources of energy are being given national priorities. Even

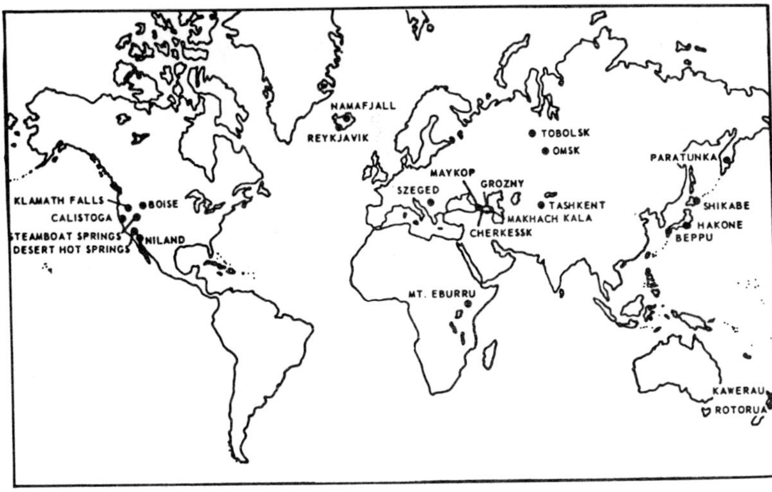

Figure 25. Regions of the world where geothermal energy is utilized for other purposes than for electricity generation.

countries with large fossil fuel reserves are investigating geothermal energy. The United States government has proposed an initial $185 million research and development program through 1980. Although this amount falls far short of any type of full scale geothermal implementation, it does grant legitimacy to geothermal energy and acknowledge its presence.

Therefore, we believe the world's research and development programs pertaining to geothermal energy will be extremely accelerated. Over the past ten years geothermal resources have been examined by many countries both on their own and through various United Nations programs, which formerly would not have considered this energy a natural resource. Many peoples can see the potentials of this energy source left dormant for the existence of the earth and in the next ten years we should see the actual implementation and utilization of this energy source.

CHAPTER 4

EXPLORATION AND DRILLING TECHNIQUES

Behold, I will stand before thee there upon the rock in Horeb; and (Moses) shalt smite the rock, and there shall come water out of it, that the people may drink . . . EXODUS 17:6

Exploration

Exploring for geothermal resources, both waters and hot rock anomalies, has become extremely technical and complex. The historical method of stumbling upon a hot spring, geyser, boiling mud pool or other surface activity is being replaced by more scientific approaches. Exploration techniques involve a variety of disciplines which incorporate geological, geochemical, and geophysical studies of a region to locate possible geothermal production sites. Figure 25A is a schematic diagram of the general approach that is undertaken in the initial investigations of a high temperature region. The distribution of surface activities (i.e. hot springs, fumaroles) of the region is examined as a complete geological analysis of the region as it pertains to a geothermal resource is conducted. This includes a complete study of the area's setrology, stratigraphy, tectonic structure and hydrology, hydrothermal alteration. A chemical analysis of its waters to determine deuterium and tritium content will lead to the knowledge of the system's hydrology and fluid circulation as well as base reservoir's temperature estimations.

Further geophysical studies to obtain crustal structure information can include seismic, gravity, magnetic and resistivity surveys. Ground surveying and aeromapping are combined with all the information obtained for a given region and a geological map is drawn highlighting the thermal reservoir and its characteristics including temperature, heat flow, circulation of the thermal fluids and crustal structures. [42, 47, 58]

Geology

Areas which were formed during the silicic volcanism periods of the late Tertiary or Quarternasy systems are the locations of nearly all (if not all) of the current geothermal regions of exploitation and investigation. [21] Orogenic zones are regions of recent mountain building and are believed to occur as a result of heat in the upper mantle melting part of the lower crust causing a convective heat flow and crust movement. [56] Consequently, the orogenic zones produce regions where forces and/or other conditions within the earth effect movements of the earth's crust in the form of faults, folds, earthquakes, or the tectonic zones.

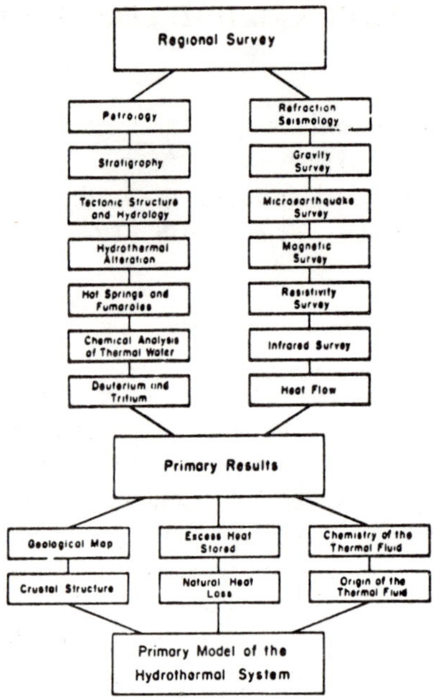

Figure 25A. Analysis of a Region for a geothermal reserve.[58]

The stresses and strains within the earth effect the upheaval of the crust causing a fold (see Figure 26). The foreland is the region of the fold which is inclined and in the direction of the thrusting forces. The hinterland is the region opposite these inclines and force directions.

Geothermal regions are generally found in the hinterlands, rather than the foreland, and these hinterland regions are generally characterized by geothermal activity including volcanism, surface thermal water activity, high temperature gradients and therefore high heat flows. Also associated with the hinterland regions are normal faults which effect fault block terrains which characterize high temperature geothermal systems. Normal fault zones can create a geothermal system in the following manner:

1. A normal fault is a fracture of the surface along an incline which shows the rocks to appear to have been pushed down with respect to rocks beneath. Normal faulting may allow for the releasing of subsurface pressures and of

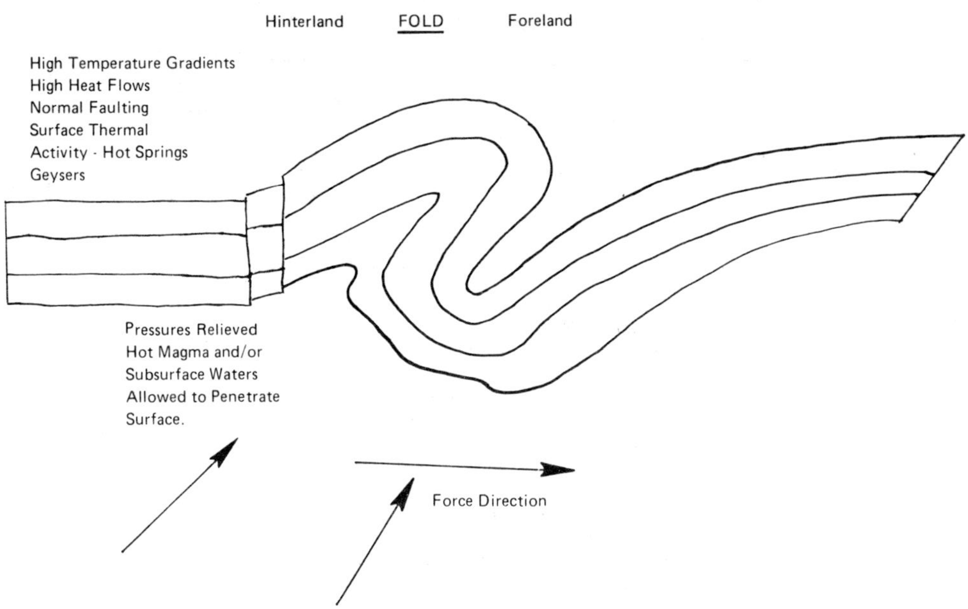

Figure 26. Geology of a geothermal system shows they are usually found in the hinterland of a fold.

rocks near their melting points, and/or allow for the passage of hot magna from within the earth to the surface.

2. Normal faulting produces a network of subterranean channels which allows subsurface waters to circulate, becomes heated, and penetrate the surface; and allows the means for surface waters to be heated and recirculated to the surface. [23, 59, 60]

Table 17 gives the average reservoir temperature that would be expected to be found with various geological formations associated with geothermal resources. Thus the geology of an area must be the primary consideration and investigations for specific geological formations are the first stages of geothermal explorations.

A regional geological survey takes place in search of geothermal formations. Surface manifestations of geothermal energy are explored (hot springs, geysers, fumaroles, boiling mud pools, CO_2 vents, steaming ground, etc.) Regions of abnormally high heat flows or temperature gradients are also noted. Regional topography and hydrology and climate characteristics are compiled and interpreted. These include rivers, lakes, average temperatures and variations, annual rainfall and seasonal variations and drainage patterns.

Table 17. The Geology of a Region is the Primary Consideration both for Geothermal System Discoveries and their Productivity.

Geologic Environment	Average Region Temperature To Be Expected. °C
1. Fault block terrains associated with Quaternary volcanism	250
2. Volcanic structures associated with Quaternary volcanism	220
3. Fault block terrains in Cenozoic hinterland regions, without Quaternary volcanism	190
4. Cenozoic Rift zones, without Quaternary volcanism	120
5. Sedimentary basins in Cenozoic hinterlands, without Quaternary volcanism	85
6. Foreland and platform regions, without Quaternary volcanism	30

Rock types and distributions are examined and geological maps are drawn usually to a scale of 1–10,000. Figure 27 shows a typical geological map. It is of the Mastukawa area in Japan and is characterized by numerous folds. Notice the faults and apparent uplifting and bending of the earth's crust.[61] There preliminary mapping and surveying procedures are for the selection of a few possible geothermal reserve sites which would warrant more intensive geophysical investigations.

Geophysical

The topography and other physical characteristics of the surface do not show entirely what lies beneath. There are regions of the earth where surface temperatures are ambient, no thermal surface activities are noted, and 4000 meters below high energy sources exist.

Geothermal reserves are covered with sediments, lavas, various other rocks and deposits, frozen earth and ice. Thicknesses may vary from a few meters to thousands of meters. Not only are more extensive geophysical methods of determining

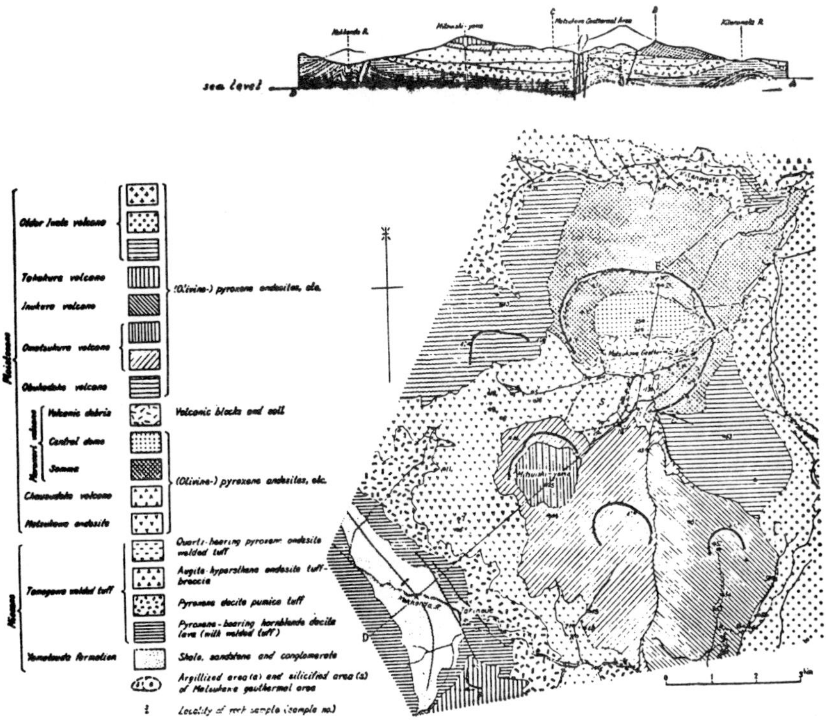

Figure 27. Geological map of the Matsukawa geothermal area, Japan.[61]

geothermal energy reserves necessary from an availability viewpoint, but also because of high costs of exploratory drilling. If an area shows some thermal surface activity and therefore the possibility of a geothermal resource below the surface, before any drillings are attempted more data must be collected to confirm the energy resource. On the other hand, new energy reserves heretofore unknown may be discovered.

Each area of investigation is different requiring different examinating techniques. Also, each government and/or company will spend a different amount of money for its geothermal research and an extensive geophysical survey can get rather expensive. Our discussion is of some of the available geophysical techniques and their applicability to geothermal energy. The combinations and variations of these techniques are many and can be applied to suit the researchers needs and budget.

From the geology and previous discussions it is seen that geothermal areas are dependent on subsurface structures, and those of importance to geophysical exploration include:

1. Fissure zones and major faults lie below geothermal systems located in sedimentary and volcanic rock encasements,
2. Secondary fissures and minor faults at shallow depths,
3. Volcanic rocks associated with geothermal systems have varying properties (i.e. magnetic, electrical resistivity). [63]

Gravimetry

Under certain geological conditions, gravimetry can identify regions of high heat flow, as a result of the metamorphism of sediments by circulating mineralized fluids. At Imperial Valley, California, there is a definite correlation between high heat flows and gravity highs. [64] Changes in gravity (i.e. rock densities) reflect major and minor fault zones beneath the surface, where they begin and end at the basement of the geothermal field. By setting a known standard density and comparing, rock types and thermal anomalies can be uncovered. Fault zone charting as well as cross sections of thermal areas including hydrothermal alterations and mineralization are also accomplished through gravimetry.

Magnetics

The magnetic method of exploration of geothermal fields is dependent on the change in magnetic properties of magnetized minerals. Volcanic rocks' magnetic properties are variable and depend on original lava flow and cooling rate, homogenity of material, surface topography location of the earth, etc. Changes in the magnetic properties of magnetic minerals are effected by hydrothermal alteration and his temperatures. As a result of hydrothermal alteration magnetite converts into non-magnetic pyrite. Magnetic minerals lose their magnetic properties when temperatures exceed the "Curie point". (The Curie point of most of the magnetized minerals is much higher than most geothermal reservoirs, but it is not known exactly what has happened over a few million years). [47, 62, 63]

Therefore, through the use of a magnometer, magnetic analomies can be located and magnetic contours of a region can be charted. Magnetic surveys can determine subsurface rock types, hydrothermal alterations and possibly hot rock deposits.

Seismic

Seismic surveys can include both reflection and refraction methods. By measuring the time seismic waves take to travel between two points various densities and crystal structures can be identified. Reflective seismic prospecting has been used to determine the best drilling site for production wells and also to clarify the subterranean picture by locating faults and fracture zones. [38] Refractive seismic studies have located underground structures and density contrasts. [63, 65]

A variation of the seismology technique is the geothermal ground-noise survey. [66] These surveys are based on the fact that the increased noise levels are associated with geothermal reservoirs. The method consists of utilizing a seismograph to measure the power spectrum of the verticle background noise of the region under study. Ground noise analomies then indicate the presence of a geothermal reservoir.

Electrical Resistivity

The electrical resistivity of rocks is highly variable, but in general it is dependent on the rock's moisture content. Rocks having no moisture have high electrical resistivity. The existence of connate waters, which usually contain various ions, form an electrolyte solution. The connate waters encased in a rock formation's pores and fissures will conduct an electric current passing through the formation. The resistivity is inversely proportional to rock bed porosity, permeability, water saturation, ion concentration, and temperature. All these factors, excluding ion concentration are indicative of a geothermal reservoir. As a result of hydrothermal alteration and weathering clayey materials are generated, and a formation's resistivity declines with clayey material content. Also, these clayey materials will tend to block passages and plug pores to form a cgsrock over a permeable formation to effect a geothermal reservoir. Thus, in the surveying of a geothermal field and the subsequent electrical resistivity testing which may involve various methods (i.e. horizontal exploration; vertical soundings); resistivity minima obtained are indications of the existence of geothermal reservoirs. [21, 47, 62, 64]

By applying Archie's formula [62, 67]

$$\rho_o/\rho_w = \Phi^{-m}$$

where
ρ_o = formation resistivity
ρ_w = water resistivity (measured directly)
Φ = porosity
m = cementation factor (estimated)

the resistivities of a number of formations can be calculated and compared to define the geothermal reservoir more acutely. Figures 28 and 29 show the relationship of the resistivity of certain rock types to porosity in water saturated rocks and temperature. Note that as porosity and thus water content (signifying possible thermal waters) and temperature increase resistivity decreases.

In the exploration of a geothermal resource electrical prospecting methods have become invaluable and therefore highly sophisticated. Prospecting for conductive zones has been accomplished through the use of direct current electrical methods, however, induction electromagnetic methods are proving to be more effective tools to the geothermologist. Methods of electromagnetic induction prospecting include:

a. audio-frequency magneto-telluric method (can be used for depths up to a few hundred meters).
b. two coil profiling method (provides high resolution at depths within 20–30 meters).
c. two coil frequency-domain sounding method (may be used for depths of 1 or 2 kilometers).
d. coil/wire time domain sounding method (may be used to map the source of anomalous heat into the lower crust and upper mantle). [67]

Since the resistivity of the rock is directly proportional to the resistivity of the trapped water in its pores, and the water is usually an electrolyte solution of NaCl, $CaCO_3$, and KCl, and the resistivity is a function of the salts concentration in solution; then salt concentrations (salinity content) of the water can be determined through the use of an electrical resistivity method. [67, 69] Therefore, as seen in Figures 28, 29 and especially Figure 30, through the use of some electrical resistivity method (and there are many and they are varied: i.e. the Wesnner Method and Dipole Method [69] use D.C. sources while those previously discussed used A.C. sources) a geothermal reservoir cannot only be discovered from the surface, but also the rock type, the temperature, and whether or not it is wet or dry can be determined. If it is wet the amount of water in the reservoir, the water temperature and the water's solution composition can be known.

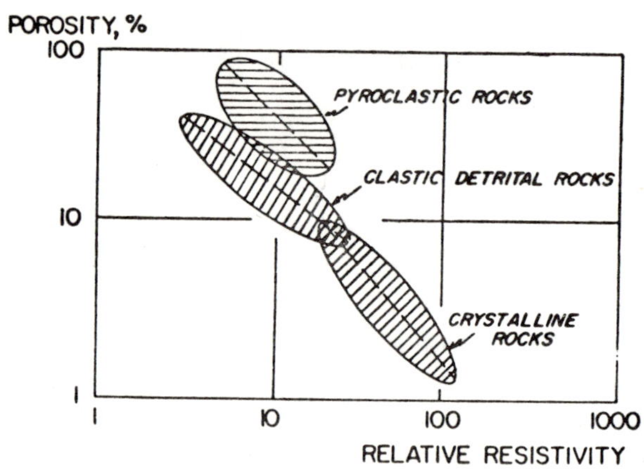

$$\text{Relative Resistivity} = \frac{\text{Bulk Resistivity of a Rock}}{\text{Resistivity of Water Contained in the Pore Spaces}}$$

Figure 28. The relationship between resistivity and porosity in various water-saturated rock types.[67]

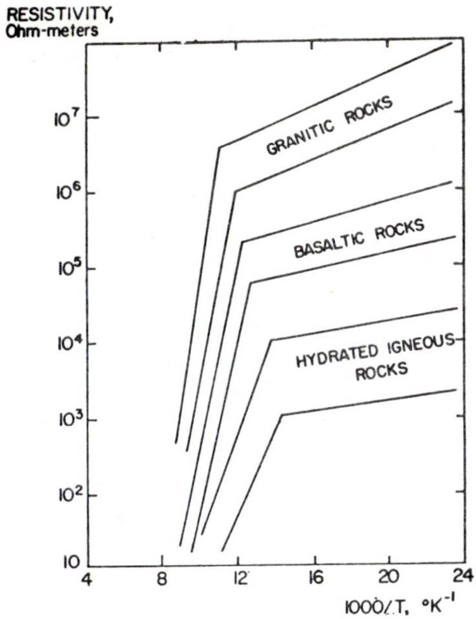

Figure 29. The resistivity of various dried rocks as a function of temperature.[68]

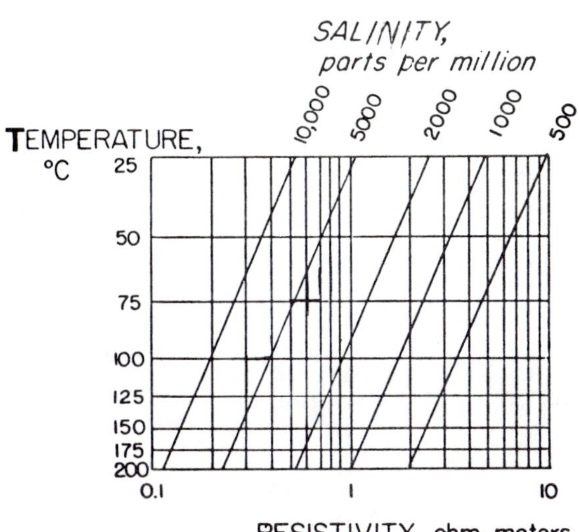

Figure 30. Resistivity of geothermal solution (NaCl) as function of temperature and salinity.[70]

Thus without exploratory drilling it can be determined through an electrical resistivity method whether or not a geothermal resource exists and if it were sufficiently large for practical applications and warranting further investigations.

Infrared and Remote Sensing

Aerial infrared surveys are used over high temperature regions and outline thermal surface patterns. The subsequent interpretation of these patterns can give insights to possible geological structures controlling the heat flow. [71]

Microwave radiometry, like infrared imagery is based on the fact that all bodies above absolute zero in temperature emit energy as electromagnetic radiation. However, where infrared imagery sensors can detect only surface effects, microwave radiometry can penetrate a few meters. Other methods such as multiband photography are even more limited in scope. The problem with these remote thermal sensing systems is that they are relatively new to geophysical exploration and their techniques have not been refined and their possibilities analyzed. [72, 73]

Through experimentation and practical use these techniques will be able to be applied to a wide range of geophysical and geothermal development, Figure 31 shows some of the possibilities for remote sensing techniques.

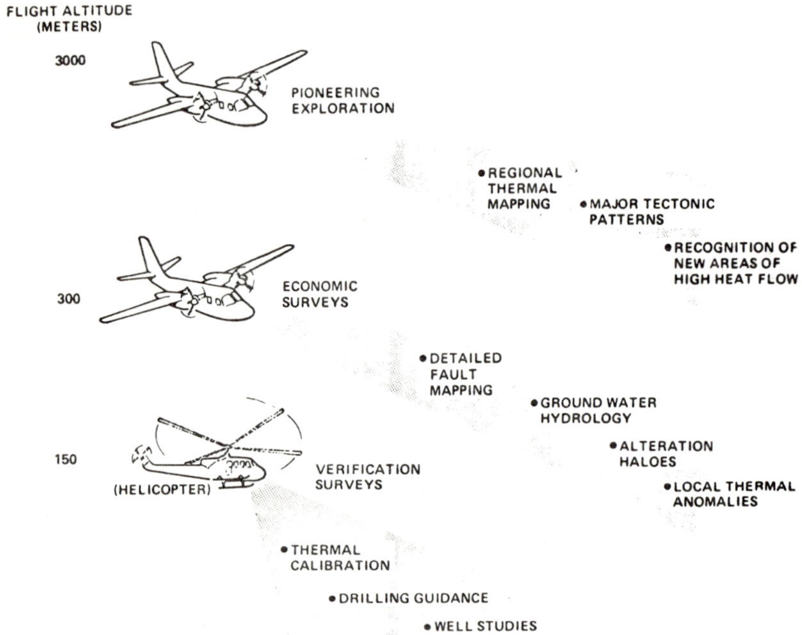

Figure 31. Eventually remote sensing techniques will be used a wide range of geothermal functions.[71]

A more extensive geophysical study might include ground surveys, micro-seismicity surveys, and/or radio-frequency interference. Heat flow measurements might be supplemented by thermal gradient drillings and various surface temperature measurements. In fact, the suggestion has been made for the measurement of the heat flows and thermal gradients of all wells in the world to locate hidden geothermal resources.

Thus, a geophysical study is the combination of a large number of research techniques. Some regions might require more extensive studies than others and a larger number of techniques will be utilized. The techniques complement each other and clarify different phases of the geothermal resource structure to produce a total picture. Costs factors are a major controlling factor to geothermal research and can limit the number of geophysical techniques exploited. However, the geothermologist is equipped with a geophysical technology that needs improvement and experience in some phases, but that is proving invaluable to geothermal research.

Geochemical

The correlation between chemical and thermal actions of the earth's crust are extremely important in the tracing of geothermal analomies. Water supply and permeability are necessary in most cases for the production of geothermal steam. Therefore, geochemical data on the fundamental water chemistry for waters classification and other analysis for boric acid, nitrites, nitrates, ammonia, lithium, rubidium, fluorine, pH determinations should be included over a large area to view the entire scope of a regions surface drainage and geothermal processes. Elements, compounds, and gases relevant to geothermal reservoirs must be determined as available in thermal waters and not in ordinary surface waters. Geochemical analyses should include:

a. fundamental water chemistry
b. identification of chemical constituents relative to reservoir temperatures
c. dissolved gas content
d. geochemical indicators of various processes associated with high temperature reservoirs

Explorations can be carried out in three phases.

1. Regional Geochemical Exploration is involved with areas of thousands of square miles. By sampling stream sediments and waters geochemists might find mercury or other water pollutants which might indicate a thermal analomy. Mercury is considered a geochemical indicator for rocks of high heat flow, especially sedimentary rocks because mercury is related to thermal analomies of high heat flow. Regional hydrogeochemical surveys are carried out for data pertaining to surface drainage, supplies, and circulation.

2. Strategical Geochemical Exploration is involved with areas less than one thousand square miles but greater than sixty square miles. This study emphasizes warm and hot spring systems including water temperatures and gas surveys. An understanding of each geothermal system and its processes is sought.
3. Tactical Geochemical Exploration is involved with areas of a few square miles. This investigation includes detailed information on the following:
   a. specific thermal springs and their surrounding area,
   b. the origin of each spring including feeding and supply systems,
   c. reservoir temperature estimations,
   d. dissolved gas content of the thermal waters as well as free gases in the systems,
   e. geochemical monitoring of the systems. [74]

Chemical Characteristics

Geochemical analysis and interpretation of surface thermal activity has lead to information concerning a geothermal reservoir:

1. temperatures
2. hydrostatic pressures
3. gas pressures
4. rock types
5. flow processes [74, 75]

There is a direct relationship between a geothermal water's chemistry and its original reservoir temperature. The elements contained in the waters whose solubilities in the waters are temperature dependent are of utmost interest. Silica, magnesium, and the ratio between sodium and potassium are the elements most considered. Hot water and rock minerals coexist at equilibrium at the lower depths. Hot waters containing various silica and/or magnesium concentrations and Na/K ratios and flowing rapidly to the surface an give accurate measurements of reservoir temperatures. [75, 76] Figure 32 shows the solubility of silica and other silica compounds in water as a function of temperature. Increasing silica concentrations mean higher temperatures at the reservoir.

Predictions of temperature through geochemical analysis has proved to be quite accurate. A comparison of temperatures estimated from silica concentrations and Na/K ratios in waters and with resistance thermometer measurements is shown in Table 18. Na/K ratio predictions were usually closer to the measured value than the silica temperatures. Another temperature indicator is magnesium. Magnesium concentrations vary with water temperatures and high water temperatures usually means low magnesium concentrations.

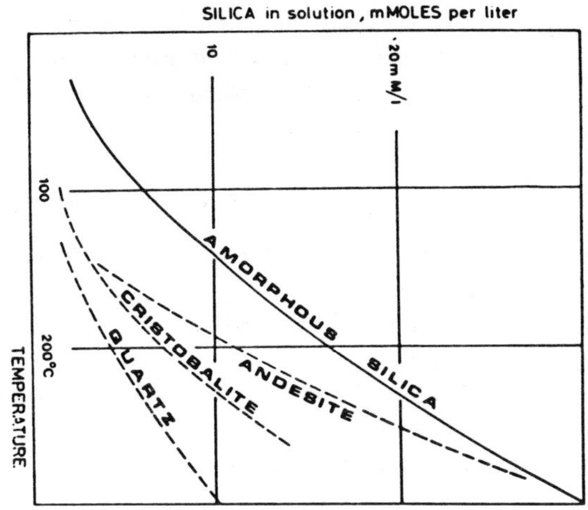

Figure 32. Silica solubility in water as a function of temperature.[77]

Table 18. A Comparative Study - Temperatures Calc. from Silica Consentrations and Na/K Ratios in Versus Direct Methods.[42]

Hole no.	Physical measurement*	Temperature °C	
		Na/K	SiO_2
1	278	230	236
2	287	278-292	262
3	281	277-288	252
4	273	268-280	260
5	244	197-200	—
6	143	192	—
7	279	252-260	270-280**
8	273	300	255.5
9	294	292	259-261
10	280	240-242	244.5
11	271	290	261
12	274	220	212
13	276	282	255
14	297	277	259.5
15	298	273	290**
16	277	234	184
17	291	298	255.5
18	290	288	240.5
19	272	292	267

* Measurements made by Ministry of Works, Wairakei.
** Results suspect or unreliable.
Silica concentrations and temperatures have been corrected for enthalpy variations.

Usually the geochemist is concerned with a large range of parameters which must be interpreted before a geothermal reservoir is identified. High temperature waters with high salinity concentrations have the lowest pH values. However, other elements present in the waters such as sulfur and calcium alter the relationships between parameters. Thus each well or geothermal area must be analyzed and the results interpreted on the basis of that particular field and its surrounding environment.

The geochemists must analyze large volumes of data including: (See Table 19).

1. water content – elements, steam, dissolved gases, minerals, heavy metals, ion concentrations, isotopes.
2. pH – as a function of the water content and its parameters and possible variation from the norm.
3. temperature – as a function of water content and other parameters.
4. phase diagrams and equilibrium constants.
5. precipitates – antimony, arsenic, sulfides, calcite, gold, silver, thallium, and mercury.

With knowledge of the above the geochemist (for a particular geothermal well site or field) can predict underground reservoir temperatures, pressures and fluid composition and constituents. He can predict CO_2 content of the reservoir's water and whether or not calcite will be deposited in pipelines. Thus, he can determine if it is economically feasible to further develop a geothermal reservoir.

Exploratory Drilling

Upon completion of the preceding exploratory techniques and the selection of a favorable site for production is made, exploratory drilling is begun. Figure 33 is a schematic summary of the investigations carried out and the results obtained through exploratory drilling. Many of the geological, geophysical and geochemical methods are further deplored during this stage.

Exploratory drilling is an important part of geothermal resource utilization for the following reasons:

1. to confirm and revise the previous information obtained on a geothermal resource,
2. to prove whether or not a site is economically and technologically feasible for full scale geothermal exploitation,
3. to provide the information necessary for the location of drill sites, for the design of production equipment and probable useful life.

Drilling

Drilling for geothermal energy is not unlike drilling for gas and oil. There are some problems of geothermal drilling; high temperatures affecting the circulatory

Table 19. Various Data and Thermal Waters Composition for some Worldwide Springs and Wells.[75]

Source and Reference	Approx. temp. (°C)	Depth (m)	pH (20°)	Concentrations in mg/kg for water collected at atmospheric pressure from discharge																			Molecular Ratios				
				Li	Na	K	Rb	Cs	Mg	Ca	Mn	Fe	F	Cl	Br	I	SO_4	As	Total CO_2	Total SiO_2	Total B	Total NH_4	Cl/SO_4	Cl/B	Cl/F	Na/Li	Na/K
Drillhole, Reykjavik Iceland (1)	100	600	8.6	< 0.1	95	1.5	< 0.02	< 0.02	—	0.5	—	—	—	31	—	—	16	—	58	155	0.03	0.1	5.2	400	—	> 300	110
Hole G-3, Hveragerdi, Iceland (1)	216	650	9.6	0.3	212	27	0.04	< 0.02	0.0	1.5	0.00	0.1	1.9	197	0.45	0.0	61	—	55	480	0.6	0.1	8.7	100	55	200	13
Pauzhetsk drillholes, Kamchatka, U.S.S.R. (2)	170 195	300-400	8.9	—	Na + K 940		—	—	7	119	—	—	—	1470	—	—	164	—	61	170	31.3	0.7	24	14.3	—	—	—
Drillhole 1, Ngawha New Zealand (3)	230	585	7.4	12.2	950	80	0.8	0.4	Ca + Mg 28		0.02	0.1	0.8	1625	—	1.3	17	—	61	460	1200	46	260	0.41	1100	23.5	20
Drillhole 2, Broadlands New Zealand (3)	260	1030	8.3	11.7	1050	210	2.2	1.7	0.1	2.2	0.009	0.25	7.3	1743	5.7	0.8	8	8.1	128	805	48.2	2.1	590	11.0	128	27	8.5
Drillhole 24, Wairakei New Zealand (3)	250	830	8.3	13.2	1250	210	2.9	2.5	0.04	12	0.015	0.05	8.4	2210	5.5	0.3	28	4.5	17	670	28.8	0.2	210	23.5	140	28.5	10.1
Drillhole 2, Rotokaua New Zealand (3)	220	880	7.8	10.2	1525	176	—	—	—	50	—	—	6.6	2675	—	0.2	120	—	55	430	102	1.2	60	8.0	217	45	14.7
Hole E103, Tahuangtsui Tatun, Taiwan (4)	200	1000	3.2	—	282	54	—	—	89	0	—	1363	—	1223	—	—	1462	—	—	170	—	·	2.26	—	—	—	8.9
Hole 1, Matsukawa, Japan (5)	~ 300	945	4.9	—	264	144	—	—	8.7	22.9	—	508	—	12.4	—	—	1780	—	26	635	61.1	—	0.019	0.062	—	—	3.1
Hole 5, Mexicali Mexico (6)*	340	1285	—	19	5820	1570	—	—	8	280	—	0.2	—	10420	14.1	3.1	0.0	—	1653	740	12.4	—	—	256	—	92	6.3
Hole 1, I.I.D. Salton Sea California (7)*	340	1600	4.7	215	50400	17500	135	14	54	28000	1400	2290	15	155000	120	18	5	12	7100	400	390	386	80000	121	5500	705	4.9

* Concentrations and pH in deep aquifer.

Total CO_2; SiO_2; etc., is the total $CO_2 + HCO_3^- + CO_3^{-2}$ expressed as CO_2; silica + silicate as SiO_2; etc.

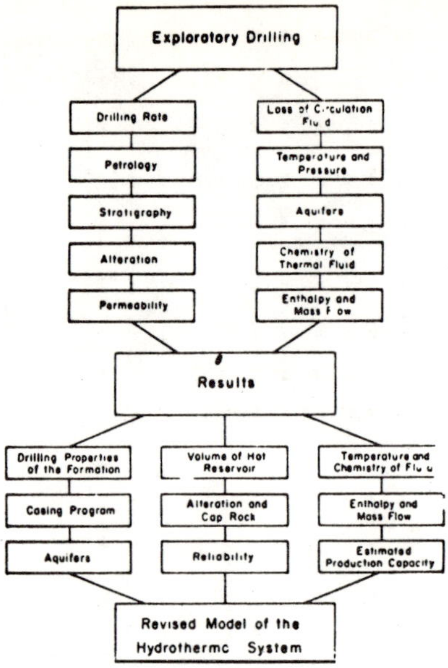

Figure 33. Schematic summary of exploratory drilling investigations and the results obtained.[58]

system, cementing procedures, casing design, drilling string; which must be satisfied to ensure safe and proper drilling.

The location of the drilling site, besides those factors pertaining to exploration, must also take into account drilling feasibility. The location should be of relatively flat terrain to support the drilling rig. Further considerations should be given to road conditions for the transportation of personnel and machinery to the site, availability of utility services such as electric power, water, and other materials for field operations and personnel usage.

Drilling Equipment

Drilling rigs for geothermal energy are usually of a medium depth capacity, between 3000 and 4000 feet, because of the relatively shallower depths of a geothermal resource than oil or gas. Figure 34 is the plan view of a typical drilling operation. The drilling machinery and equipment may be for a mechanically driven operation, although electrical drive is optional depending on economics and location feasibility. Figure 35 shows another view of a drilling rig and some typical drilling times for various depths of wells.[78, 79, 80] The number of days required to drill may vary from these times by over 100 percent depending on geological formations and other conditions. [80, 81] Included with the drilling apparatus is usually a mud cooling tower attached to the mud circulatory system.

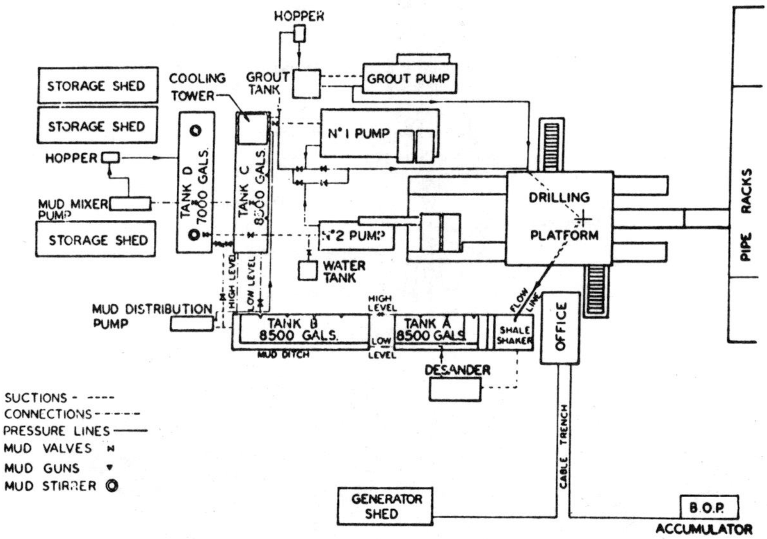

Figure 34. Equipment layout and flow lines of a drilling rig for geothermal energy.[78]

Obviously each geothermal well will be different with regards to rock formations, steam quantity, steam quality, corrosive constituents and other factors. Therefore, the following discussion will try to give the reader a feeling for geothermal drilling with typical values for a typical well operation.

Hole Size and Casing Design

Depending on anticipated and actual steam and/or hot water volumes of a well site, a hole size is selected. Table 20 shows the various hole size to casing design with respect to steam volume.

The purpose of the casing is to allow passage of large quantities of steam from the underground geothermal reservoir to the surface. The casing prevents or inhibits hole wall collapsing, steam contamination and hot water flooding of the hole from above the reservoir. The casing should be able to withstand vibrations, rolling, friction, wear and corrosion.[80, 81]

Rolling and vibration are inhibited by designing the outside diameter of the casing pipe to be nearly equal to the hole size. The inside of the casing is worn as a

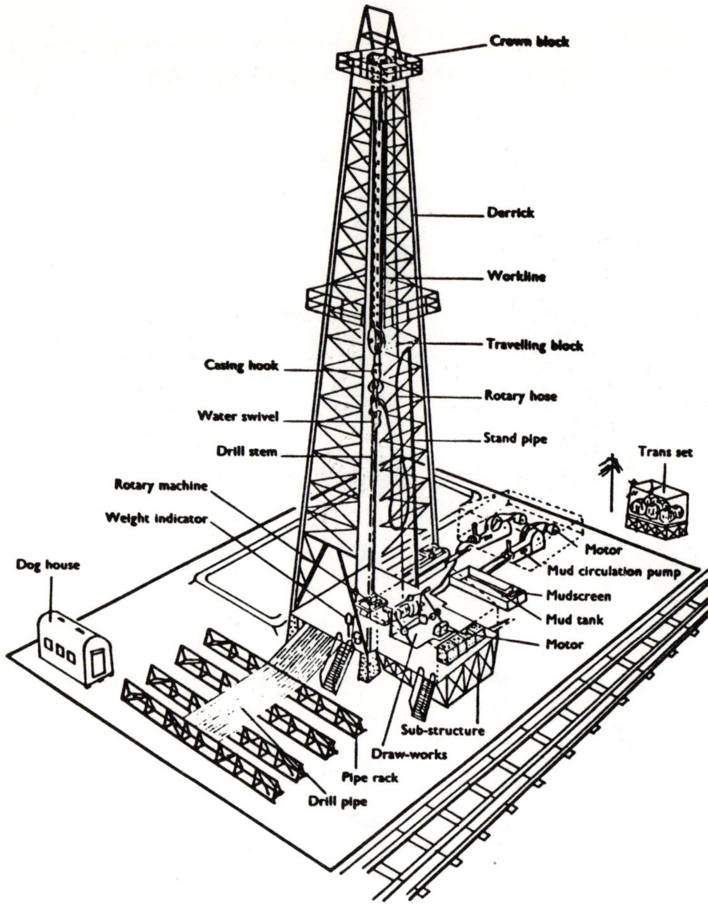

Depth	Actual drilling	Finishing and testing
500 m	15 to 30 days	10 days
1,000 m	25 to 45 days	10 days
1,500 m	35 to 55 days	10 days
2,000 m	50 to 70 days	10 days

Figure 35. Typical geothermal energy drilling rig assemblage and typical drilling times.[80]

result of high speed wet steam, improperly fitted or designed joints, turbulent steam flow. Corrosion of the casing is effected by iron sulfide formation (casing reaction with hydrogen sulfide), and high temperature induced stresses. Proper material selection and design criteria will help alleviate most of these problems.

Table 20. Hole Size and Casing Design as a Function of Steam.[50]

Open hole	Casing	Cementing
(a) Steam volume 10 to 25 t/h:		
17-in open hole	13 3/8″ surface casing	fully cemented
12¼-in open hole	9 5/8″ intermed: casing	fully cemented
8 5/8-in open hole	7″ production casing	fully cemented
6¼-in open hole	4½″ slotted liner	
(b) Steam volume 25 to 50 t/h:		
18-in open hole	16″ surface casing	fully cemented
14¾-in open hole	11¾″ intermed: casing	fully cemented
10 5/8-in open hole	8 5/8″ production casing	fully cemented
7 5/8-in open hole	6 5/8″ slotted liner (O.D. of coupling skimmed by 1/16″	
(c) Steam volume 50 to 80 t/h:		
22-in open hole	18″ surface casing	fully cemented
17-in open hole	13 3/8″ intermed: casing	fully cemented
12¼-in open hole	9 5/8″ production casing	fully cemented
8 5/8-in open hole	7″ slotted liner	

Casing Cementing

A cement sheath covering the entire length of the casing is provided for to minimize vibrational and frictional and thermal expansion wear. Casing cementing for a geothermal well is different from an oil well in that high temperature cement must be used otherwise all other processes are essentially the same.

Circulating Drilling Mud System

The drilling mud should have the following features to ensure good drilling operations:

1. Ability to break up the rock drilled by the rock bit;
2. ability to remove drill cuttings from the well bottom to the surface;
3. ability to waterproof the walls of the well with a thin, non-permeable mud cake to prevent caving;
4. ability to cool and lubricate bit, pipes and formation;
5. ability to control fluid pressure in the formation (prevent water entrance) through mud water column pressure;
6. no characteristic changes as a result of dissolved substances of formation, cement, salt water and high temperatures.[81, 82]

The circulating fluid medium of Imperial Valley, California is fresh water gel-lignite mud. The legnite counteracts the effects of high temperatures by preventing clay particle flocculation and thus acts as a thinning agent in the mud. Problems arise from lost circulatory fluids which cause delays and extremely high temperatures which require cooling towers to lower circulating temperatures from about 200°F to about 20° to 30°F.[83]

Wellhead Equipment

The system of piping, fittings, valves, and connections comprise the special wellhead installed at the surface of the geothermal well. Wellheads for geothermal wells must be made to meet specifications to handle, control, and withstand:

1. large flow rates;
2. mud and steam of high temperatures;
3. well operating and closing pressures which are relatively low;
4. highly corrosive actions, both chemical and mechanical, of the steam, hot water and other mixtures delivered;
5. the connection of the wellhead system to large diameter pipelines.[79]

An important aspect of the wellhead equipment is safety and drilling precautions. Improper drilling techniques and lax safety precautions can result in a blowout due to extreme pressures and temperatures. In fact, a blowout at The Geysers field in California blew off the top of a hill and has yet to be capped.

The installation of a blowout preventer is imperative to the drilling operation, Figure 36 is a blowout preventer designed specifically for geothermal well drilling. It is fitted onto the wellhead as a sort of complex safety valve to prevent an uncontrolled blowout by closure on drill pipe, drill collars, casing or of just the central bore.[78,79]

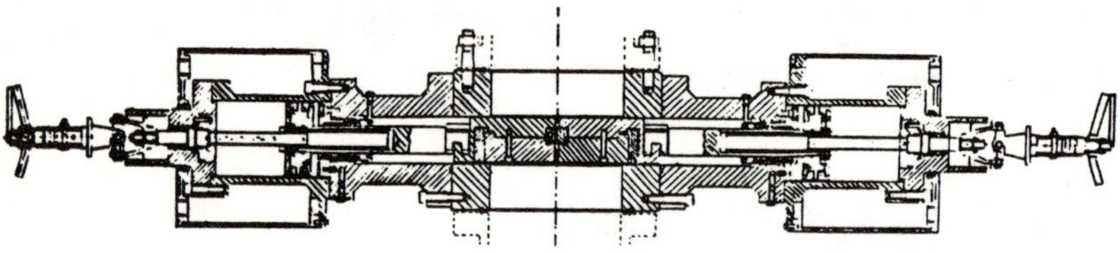

High Temperature, High Pressure and Corrosive Resistant Materials Must Be Used in its Manufacture.

Figure 36. Blowout preventer designed specifically of geothermal well drilling.[79]

Drilling the Well

Rock formations are usually as hard at the surface as they are at great depths. Therefore, the extremely hard and abrasive rock causes the loss of many drill bits. To clear the hole, a second-stage steel mill-tooth bit is required and drill collar weights of 1000–2000 lbs. per inch of bit or all available weight is necessary to drill the surface hole. A rotary speed of 70–80 rpm or a speed which will allow the smooth operation of the bit is another requirement.

As previously noted, a geothermal energy formation is usually faulted and fractured and unless the bit weight and speed is properly coordinated much bouncing and jumping will occur with many more lost bits. Therefore, drilling rates compromise to save bits to the point of economic feasibility. Further complex faulting and fracturing may result in direction drilling along the path of a natural rift. This is one stage which emphasizes the importance of a complex geophysical investigation. However, directional drilling is very expensive and can double the cost of a drilling operation. Of imminent importance to a drilling program of this type are qualified personnel to collect, analyze, and report data and order changes in drilling operations quickly and efficiently.[83, 84] Failures associated with geothermal drilling include:

1) drilling column sticking due to:
   a) steam or gas blow-outs
   b) rapid loss of mud circulation
2) drilling at high temperatures in cold weather causing expansion of the pipe joints and separation of the tool joint from the drilling column.
3) drill collar breaking off due to excessive shocking and turbulent drilling.[80]

The above failures can be prevented and/or remedied through a variety of special lubricating and drilling techniques. Proper material selection to withstand extreme temperatures, pressures and loadings will lessen failure possibilities.

Alternate methods of drilling such as air drilling,[80, 81, 85] turbo-drilling,[81] and water and fluids[85] be considered when conventional drilling techniques are not sufficient or when they prove advantageous or economically feasible.

Completion

Continuous surveying during and after drilling is effected to ascertain proper drilling conditions and well operation. Table 21 shows the data that is collected for what purpose it is used.

Table 21. Continuous Well Surveys' Collection Data.[80]

Survey in the well	Surveying instrument	Purpose of investigation
Temp. measurement	Point survey: (thermometer)	Study of the change in temperature of the well, distribution condition of geothermal heat
	Multiple survey (Kuster), continuous survey (Thermistor)	Presumption of steam, and hot water layer
Electrical log		Comparison of formation and supplementary analysis of ground surface electric detection
Measurement of flow inside well	Spinner	Measurement of flow of steam
Inclination measurement	Totoko and Murata Type	Measurement of direction and inclination of the well and presumption of bottom hole position of the well.
Pressure measurement	Amerada (Bourdon tube)	Measurement of the pressure in the well
	Humble type (Spring)	Presumption of production capacity
Collection of bottom hole sample	Bottom hole sampler	Collection of water under pressure in the well
Cement bond log	Improvement of sonic log	Decision on satisfaction of cementing

The new well is flushed through use of several techniques which may or may not be necessary depending on the initial well blow-out, (controlled blowouts are safe). Further testing along the above lines can determine a well's future productivity. If these tests are affirmative the well is capped and vented, drill equipment removed and production equipment moved onto the well site. Negative production results warrant the re-examination of geological, geophysical, and geochemical as well as drilling's data to verify well site selection and possible alternate means to boost production such as drilling deeper or casing perforation.

Dry Geothermal Sources

Drilling to utilize a dry geothermal source is not unlike drilling for a wet resource. Where initially steam blowout problems are not encountered, eventually the hot fluids circulated artificially will cause the same corrosion, high temperature and high pressure problems as naturally wet systems.

Dry drilling problems will be greater in some respects. Thermal shocking of the casing due to cold water initially injected into the well to reach temperatures over 250°C at the well bottom. Reservoir size must be determined so that estimated of circulatory fluid quantities can be made. Tapping of the artificial reservoir requires accurate measurements and well site selection. However, if the same hole is to be used for cold fluid injection and hot fluid extraction, other factors must be considered including difficulties of drilling large diameter holes, proper fluid circulation for maximum heat exchanging and thermal insulation between rising and falling fluids.

CHAPTER 5

GEOTHERMAL POWER PRODUCTION SYSTEMS

**"As for the earth, out of it cometh bread:
and under it is turned up as it were fire." . . . JOB 28:5**

The utilization of the geothermal resource is rather unique as compared to other energy sources and many difficulties are associated with it. Each geothermal reservoir is different from the next with respect to geological formations, thickness, hardness, permeability of the bed rock, chemical, mineral, dissolved solids and gases content of the subsurface fluids, temperatures, pressures, hot water, wet stream, dry steam, etc.

The common denominator for each geothermal reservoir is heat. Heat supplied in different quantities and different forms. However, whatever the form the heat dissipates quickly upon arrival at the surface and therefore hot fluids of a geothermal nature can not be conveyed across the earth's surface for long distances without great energy losses and lower heat qualities. Depending on the application of the heat source these transportation heat losses are magnified many times. Hot fluids in Iceland are transported many miles from the geotherml well to the capital city, Reykjavik with no appreciable loss in heating values. Steam and hot fluids to be used for the generation of electricity, which requires high initial enthalpy values of the steam, cannot be transported great distances without great losses of efficiency. Furthermore, piping and insulation costs have become so overbearing that some geothermal developers have considered using a series of turbines over an entire field to reduce piping costs and heat losses. Another problem is that the known geothermal reserves of the world and especially Western United States are not readily available to the majority of the population centers.

To reach these metropolitan areas, the heat energy of the geothermal well is converted to electricity which travels along transmission lines to its destination. While electricity is relatively easily handled, it does represent a heat loss upon conversion from heat to electricity, upon distribution along transmission lines, upon the actual usage of it for lighting, heating and other work. Because of its inability to be stored electrical power plants are designed to meet demands based on a hot summer day which seems to drain the earth of all available power. This is a drain on all resources and a problem right now for geothermal power plants for they are not as flexible as the fossil fuel plants.

Most of the problems associated with geothermal electricity are the result of a retarded technology base which should broaden as research and implementation increase. The idea of using geothermal heat for electricity production and other processes is actually quite simple. Figure 37 shows a geothermal deep well heat pump designed by the Sperry Rand Corp. The steam rises up through the well casing to the surface where it passes through a heat converter. In electrical production, the heat converter is more or less a standard power plant without its boiler and probably included in the flow schematic is a filtering system to remove corrosive and other steam constituents. The waste water from the power plant can then be reinjected into the ground to promote a continuous cycle.

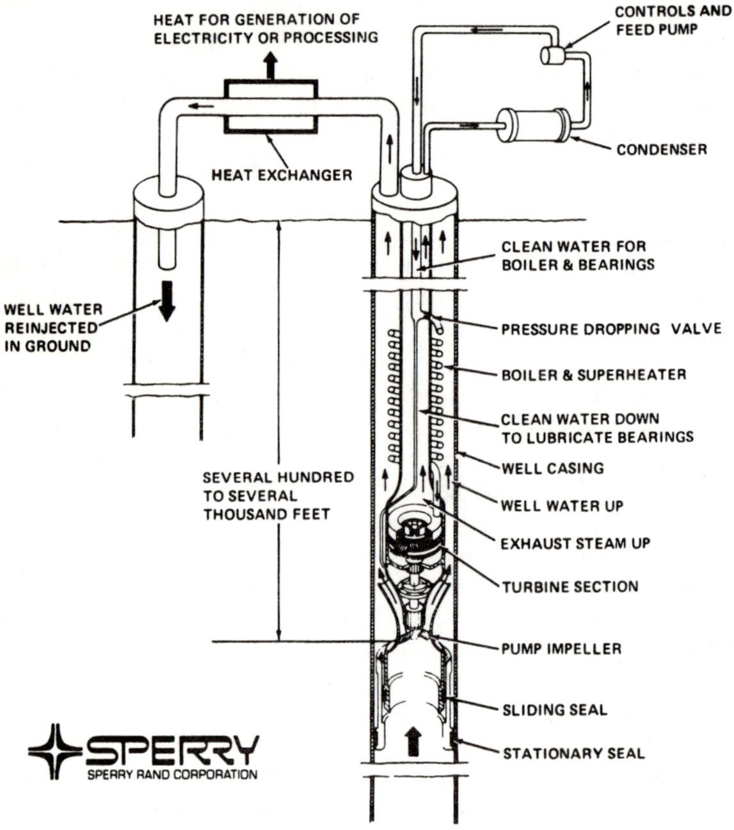

Figure 37. Geothermal deep well heat pump.[86]

Dry Steam

The dry steam producing geothermal sources of the world are quite rare. However, they do produce a majority of the electricity generated from geothermal sources. The three dry steam electricity generating areas of the world are located at The Geysers, California; Larderello, Italy; and Matsukawa, Japan. The geological characteristics of a vapor-dominated system have been discussed previously and it has been noted that these systems are characterized by dry and superheated steam existing under uniform temperatures and pressures which are lower than comparable depths of other water dominated geothermal reservoirs.

Temperatures and pressures of these systems produce a continuous vapor phase and vary with depth. Figure 38 is a plot of temperature and pressure as a function of depth. Temperatures of a vapor dominated system usually hover around 240°C with pressures over 450 psi. Variations from the norm can be expected when the steam is contaminated with dissolved solids and gases. Increased depths and corresponding temperature and pressure increases and the boiling of saline solution (sometimes referred to as water brine) results in superheated steam. A 25 percent saline solution will produce superheated steam by 12°C when boiled at 35 Kg/cm^2 [87, 91]

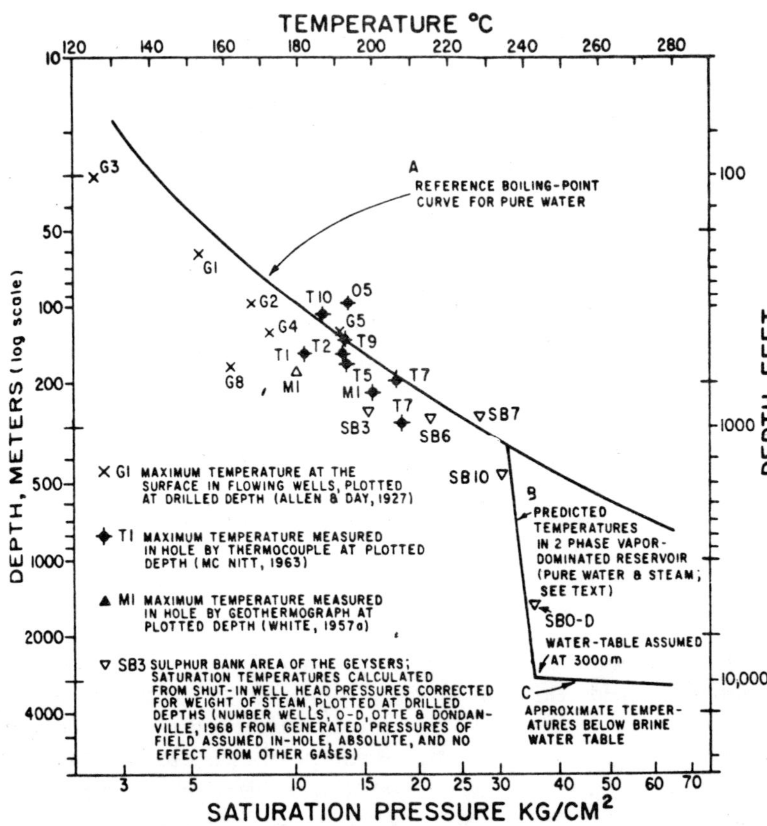

Figure 38. Temperature and pressure of a vapor-dominated system as a function of depth.[87, 88, 89, 90, 91]

Dry Steam Electrical Systems

Steam of relatively low pressures and temperatures is used for geothermal power generation making the handling of the steam relatively easy. Several wells are tapped in a geothermal field to provide steam to one geothermal power plant. Each well is in effect an individual boiler producing steam of individual characteristics. For example, six wells contribute steam to the Matsukawa Geothermal plant, Japan and all produce steam at different temperatures: well No. 1, 153°C; No. 2, 170°C; No. 5, 190°C; No. 6, 183°C.

However, when these steams are regulated and combined they are used directly from the earth and can readily drive turbines for electricity production. Figure 39 is a simple schematic of a geothermal power station using natural steam. The actual flow diagram of the Matsukawa geothermal power plant is shown in Figure 40. This station produced 20,000 kw of electricity with a condensing turbine generator.

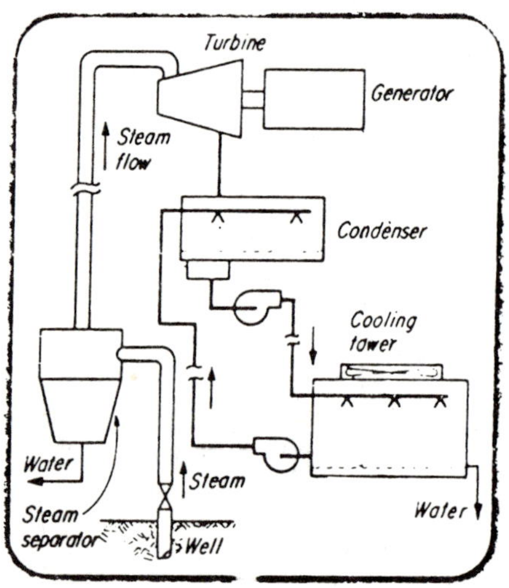

Figure 39. Simple schematic of a geothermal power station.[20]

Figure 41 is the power cycle at one of the 53,000 kw generating units at The Geysers geothermal field. This unit is typical of natural dry steam generating stations with variations occuring in flow rates and contents and equipment arrangement. At The Geysers about ten steam wells contribute steam to each unit. Steam lines from each well connect to a 36 inch steam main line which delivers the steam to the unit generating station. Installed in each well steam line are centrifugal separators which inertially remove from the steam rocks and large dust particles that have been carried up from te well. Pressure in the main steam line is regulated to 150 psig by pressure relief valves venting excess steam to the atmosphere through

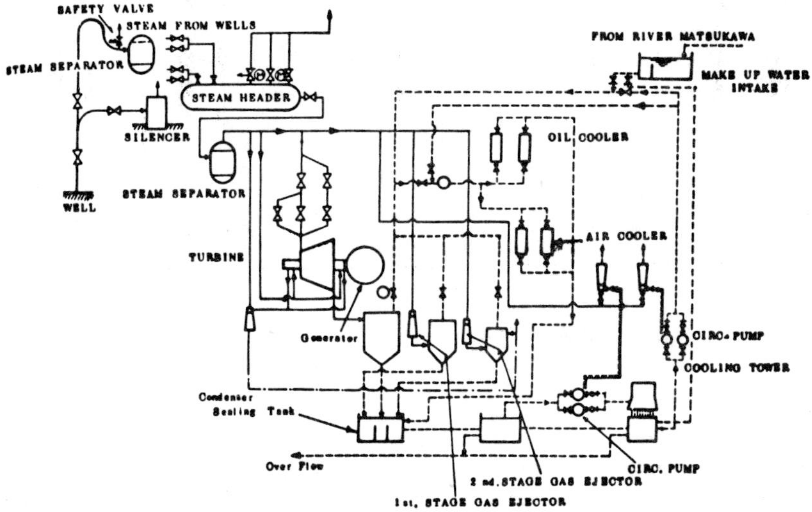

Figure 40. Flow diagram of the Matsukawa generating power plant.[92]

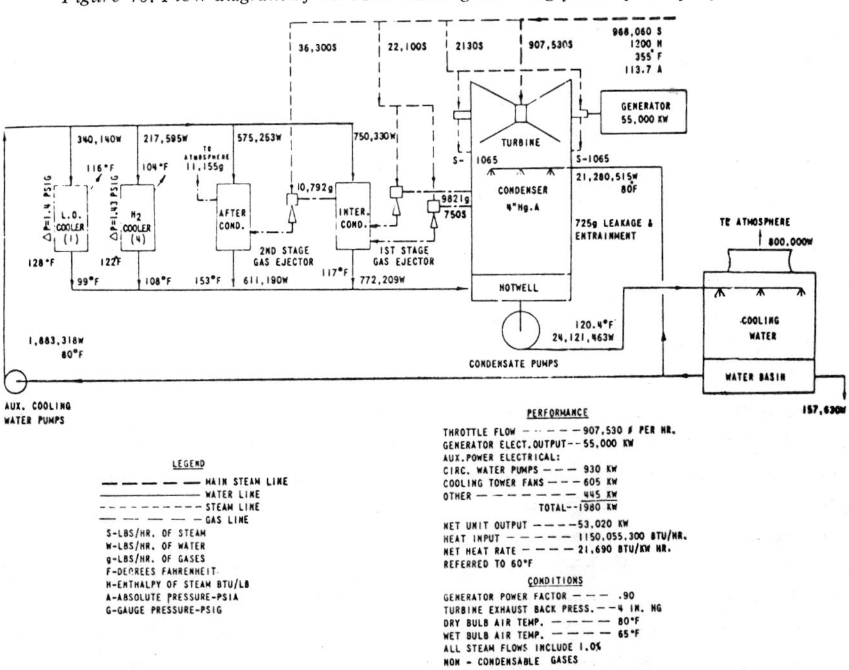

Figure 41. Geothermal power cycle at The Geysers.[31]

silencers which muffle the sound of the exiting steam. Steam conditions at the turbines are about 100 psig at 355°F. Some typical inlet steam characteristics are given by Table 22.

Table 22. Steam inlet conditions for the turbine-generator at The Geysers.[13]

Unit No.	Rating kW	Steam Flow lb/h	Steam Pressure psig	Temperature °F
1	12,500	240,000	93.9	348
2	13,750	255,475	78.9	342
3 & 4	27,500	509,600	78.9	342
5-10	55,000	907,530	113.7	355 *
11	110,000	1,808,000	113.7	355

The main steam line divides into two 24 inch lines. Y-strainers are installed in these lines to remove those larger particles which escaped the centrifugal separator. The steam then passes through emergency shut-off valves. A double set of valves has become necessary because of the suspended dust in the steam clogging valves and reducing their reliability.[93] Dust problems are characteristic of a dry steam geothermal plant. The Matsukawa power plant has had its turbines damaged by dust.[92, 93] Turbine load and speed at startup are regulated by controlling steam flow with valves. The valving again can be adversely affected by the dust.

Steam leaving the turbine passes to the condenser at 4 inches Hg absolute. At The Geysers the condenser is of the direct contact type, however, other condenser designs are being used in Japan and Italy. Cooling water falling over a series of trays mixes with the steam to condense it. Steam jet ejectors remove those gases from the condenser which are noncondensable.

The condensate and cooling water are pumped from the condenser sealing tank through the cooling tower. The cooling tower, of induced mechanical draft design, cools the hot water to about 25°C and it returns to the condenser to complete the cycle. About 20 percent of the steam condensate which enters the cooling tower is not evaporated causing a surplus of water for the system. This water surplus is re-injected into the geothermal field of origin (see Figure 37) for recycling and ultimately to extend the productivity of the steam field.[13, 31, 35, 92, 93]

In summary, Figure 42 is presented to show the various steam cycles in use for electrical energy production. Diagram 1 is for direct utilization of the steam where incondensable gases present in the steam range over 50 percent. The plant is very simple but produces less than 40 percent of the power as does the cycle in Diagram 3 under similar steam conditions.

Flow diagram 2 is a steam cycle where the steam is first purged of noncondensable gases before the turbine. This plant cycle is usually used when noncondensable gases in the steam range between 10 percent and 50 percent. The power generated from this cycle is about 60 percent less than the cycle shown in Diagram 3 under comparable steam conditions.

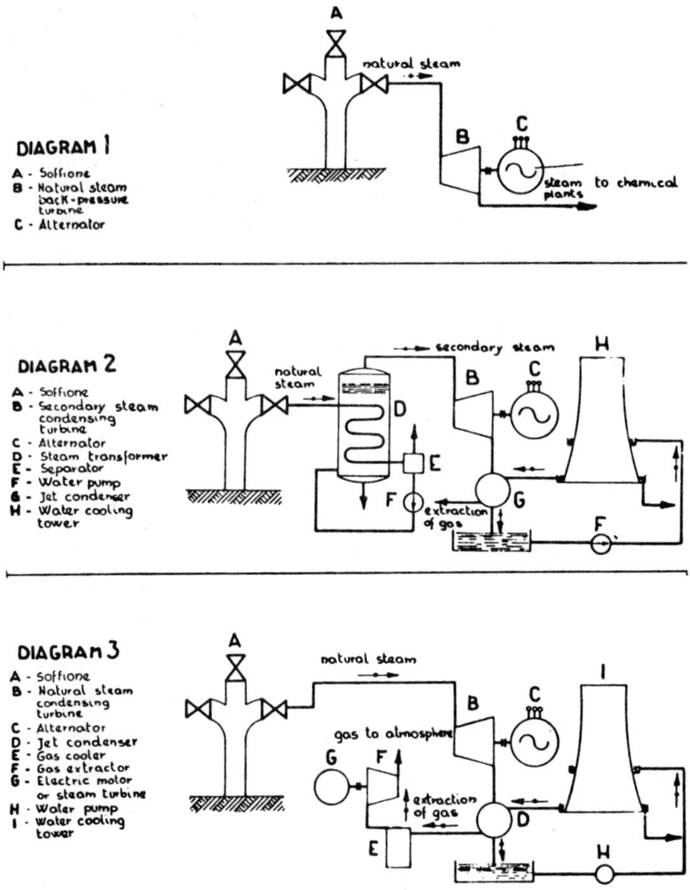

Figure 42. Flow diagrams for the various steam cycles used for geothermal electricity production.[94, 96]

Flow diagram 3 is the steam cycle we have previoulsy discussed. It is the most efficient cycle by far and is usually used when noncondensable gas content of the steam is less than 10 percent. Plants of this type are in operation at Larderello, Italy, The Geysers, California, and Japan.[94, 95, 96]

Wet Steam Systems

In Wairakei, New Zealand the first wet steam field was developed and is producing 192 Mw of electricity under the auspices of the New Zealand Electricity

Department. Other wet steam fields in production today are located at Kawerau, New Zealand; Otake, Japan; Pauzhetska and Parantunka, U.S.S.R.; Namofjall, Iceland; and Cerro Prieto, Mexico. From this distribution of power plants it becomes obvious that the wet steam geothermal resources is more readily available worldwide than dry steam phenomena.

Wet steam fields produce steam quantities comparable to dry steam fields, however, they also produce hot water in quantities that are sometimes three times the weight of the steam produces. Thus wet steam fields' electrical production differs from dry steam fields in that they only have another heat source which must be utilized. In fact, the heat from the hot water may represent 40 percent or more of the heat output of a well. Table 23 compares the steam flow characteristics and water flow characteristics with respect to quantities obtainable and heat content of each. This is data from a Wairakei geothermal field for a high pressure bore (180 psig) and intermediate pressure bore (97 psig). In both cases, the hot water exhibits heat and mas properties which demanded utilization.

Table 23. Steam and Water Heat and Mass Characteristics at High and Intermediate Pressure Wells.[97]

	H. P. Bores		*I. P. Bores*	
	Mass $\times 10^3$lb	Heat $\times 10^6$BTU	Mass $\times 10^3$lb	Heat $\times 10^6$BTU
Steam	1398	1679	1240	1479
Water	8750	3090	4400	1350
Total	10148	4769	5640	2829

Power can be generated from the hot water in a power plant similar to a dry steam plant by using steam which has been separated from the hot water by flashing it to a lower pressure. The hot waters are most economically flashed at inlet turbine pressures. However, depending on the water's initial heat properties, it may be double flashed. Double flashed steam has had steam and water separation at a certain lower pressure. Flashing pressures are dependent on each well's and each plant's operating conditions. Double flashed waters has a temperature around 100°C and has given up 30 percent of its quantity to steam. Most common plants flash about 10 percent of the hot water to steam and try to utilize the discarded water for other applications (see Chapter 6).

In fact, cycle efficiencies, like the one shown in Figure 43 for Cerro Prieto, Mexico, are about 10 percent to 15 percent; but the application of the hot discarded water for other uses can raise the efficiencies of a geothermal power plant to nearly 30 percent. The flashed steam cycle, Figure 43, differs from the previoulsy discussed dry steam cycle only because of the addition of the centrifugal separators used to separate the steam for the hot water. Figure 44 shows some typical interflash and wellhead equipment at a well site.

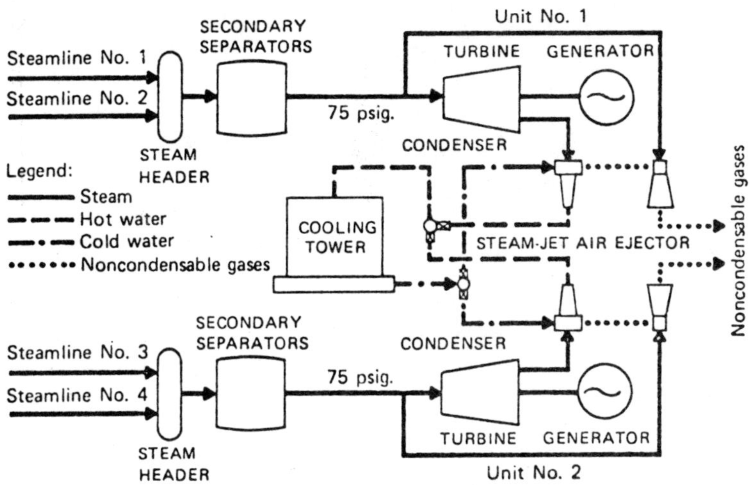

Figure 43. Flashed steam-hot water cycle at Cerro Prieto, Mexico.[98]

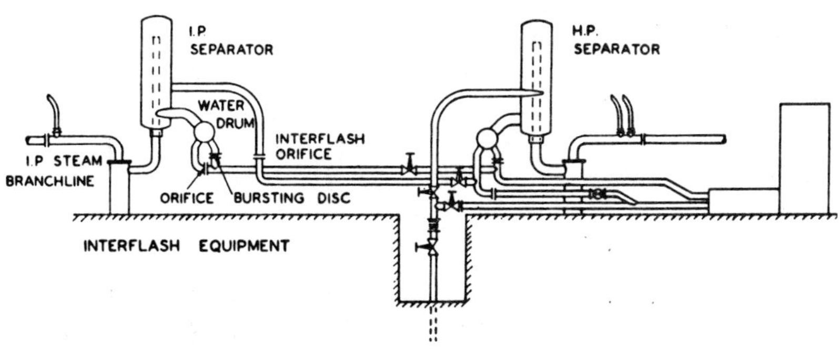

Figure 44. Typical interflash and wellhead equipment.[97]

Low efficiences for geothermal power cycles cited above are typical for the industry. The factor most contributing to these low efficiences is the vast quantities of fluids that must be circulated which results in plant energy consumption levels approaching 35 percent. Thus, total utilization of a plant's resources must become reality to increase efficiencies and the economics of a geothermal power plant.[16, 97, 98]

Hot Water Heat Exchanger Systems

The flashing production of steam becomes impractical at hot water temperature ranges between 212°F and 380°F. At these relatively low temperatures the disadvantages of steam flashing become acute:

1) Steam flashed from geothermal hot water is likely to be corrosive.

2) High water consumption and low cycle efficiencies.

3) Costlier steam turbines due to low temperature steam.

Therefore, the use of secondary fluids has created interest as a viable substitute to flashing steam. The hot water passes through a heat exchanger and transfers its heat to a liquid such as freon, isobutane, isopentane, or some fluorinated hydrocarbon, which in turn runs the power plant.

Figure 45 is a geothermal power plant which runs on a hot water-secondary fluid cycle. Hot water with a temperature over 300°F is pumped from the geothermal wells. Upon reaching the surface the water boils and is sent through a heat exchanger where it super heats isobutane to 300°F at 500 psia. This superheated vapor is at an adequate pressure to drive a turbine and therefore is expanded through a three stage radial flow turbine. The vapor then passes through a condenser and it is condensed at about 85°F by cooling water from the cooling tower. The liquid isobutane is then pumped to the heating and boiling heat exchangers to begin a new cycle. Cooling waters can also be supplied by rivers, ponds and shallow wells. Isobutane was used in this cycle designed by Magna Energy, Inc., but other secondary driving fluids are being used worldwide.

The advantages of this type of system over a flashing system are many:

1) Turbine sites are smaller, and in some cases the turbine diameter is 4 times less than the diameter of a comparable steam turbine.

2) An isobutane turbine operates above atmospheric pressure preventing corrosion due to air and oxygen leakage.

3) Low cost alloys can be used in the turbine construction for isobutane is relatively non-corrosive.

4) High efficiency turbines can be incorporated into the power plant.

5) Cycle efficiency is higher as a direct result of higher turbine efficiencies.

6) Stress and vibration reductions in the turbine, cavitation reductions in pumps and water erosion elimination of the turbine blades.

The disadvantages would include:

1) High cost extensive heat-exchange equipment would be necessary;

2) Secondary drive fluids would cost much more than water, therefore:

3) Leaks from the closed system would have to be eliminated. This cycle is available for implementation into a system for all major technical problems have been resolved and the economics of the system prove feasible.

A variation of the above cycle has been proposed as a dual cycle system incorporating a flashing steam from water cycle and hot water-secondary fluid cycle. Hot water about 500°F pumped from a geothermal well passes through two heat ex-

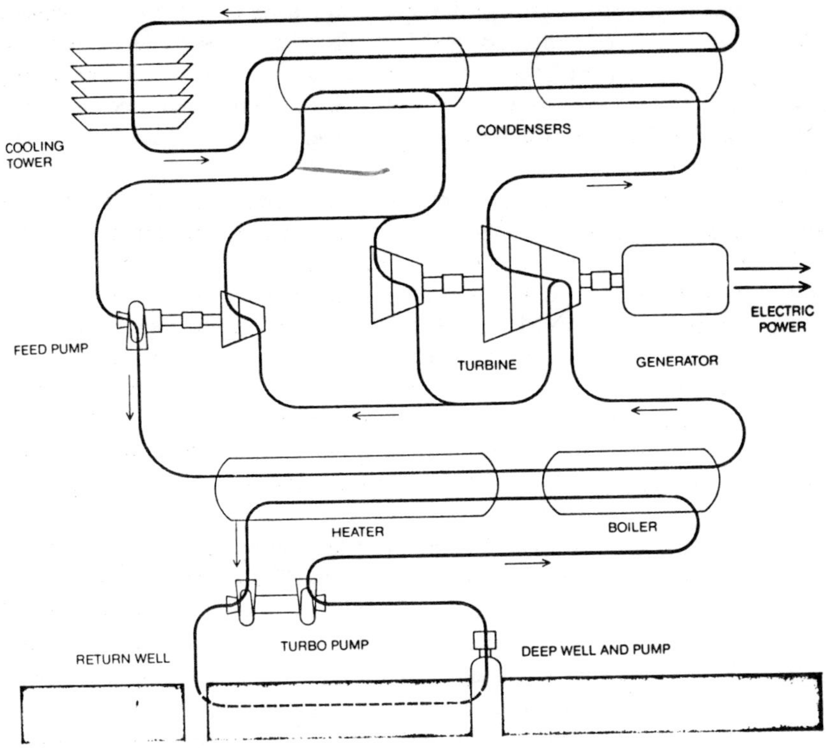

Figure 45. Geothermal power plant using a hot water-secondary fluid power cycle.[6]

changers in series and then is pumped back into the earth. From the first heat exchanger steam would be produced in a closed system (comparable to Figure 45) to drive a turbine. The second heat exchanger would have waters entering it at much lower temperatures than 500°F. It would utilize isobutane in a cycle system (similar to the one just discussed) to remove the remaining heat from the water and further power generation can result. This dual cycle system is just one of the many possibilities that can become attractive if work on them and the resultant technologies are increased. With the recycling of the hot waters back into the earth, these closed systems will produce electricity with no pollution to the air or waters.[6, 20, 98, 99]

Hot Rock (Dry Geothermal Sources) Systems

Since the vast majority of the geothermal heat resources of the world are located in dry hot rock sources rather than wet systems, it is only natural for this energy source to be tapped. Experimental work at the Los Almos Scientific Laboratory in northern New Mexico has studying the hot rock geothermal resource and methods to extract heat and energy from this energy source. In Chapter 1 we discussed some of these methods.

Figure 46 is the schematic diagram of the experimental Los Alamos system. A hydraulic fracture network is to be created approximately 3000 meters in diameter. After the initial water is pumped into this undergraound heat sink, the system will operate more or less as a closed loop with make-up water added because of underground and other water losses. Water at 65°C and 1000 psia is pumped into the earth and circulated at 7500 feet where temperatures range between 260° and 320°C and pressures of 4100 psia can be expected. The water is then pumped out of the ground and when it reaches the surface its temperature is 230°C at 1250 psia. For this experiment the hot water is to be circulated through a 100 mw air cooled heat exchanger with the extracted heat dissipated to the atmosphere.[16, 100, 101]

Testing at this site is expected to proceed for another 2 to 3 years before the actual feasibility of the program can be determined. The system is to be studied for its various geophysical, geochemical and geological characteristics over a long time period to note changes in circulation and declines in heat supplies. It is thought that in order for the heat supply to be continuous new thermal cracks must be propagated throughout the thermal crack network to expose new heat-transfer surface area. Problems that may arise are believed to be solvable through engineering techniques currently being devised.

However the final system is designed, whether a closed loop system as in Figure 46 with two pipes, a system composed of concentric pipes Figure 47, or a simple open system as shown in Figure 48, it will produce hot water. Depending on the water temperatures and pressures, the various power cycles noted above can be used. If hot enough, the water might have steam flashed from it to drive a steam turbine; if not, the waters could be passed through a heat exchanger to operate a turbine via a secondary drive fluid. (see Figures 42, 43, 44, and 45) Economics are very important and will be one of the major controlling factors of the type of power cycle finally decided upon for each hot rock geothermal power plant. The non-polluting aspects of a nearly entirely closed system will be instrumental in furthering dry hot rock geothermal technology.

Combining Resources

The many and varied applications of a geothermal resource will be discussed in Chapter 6, however, the economics of geothermal electricity production warrant

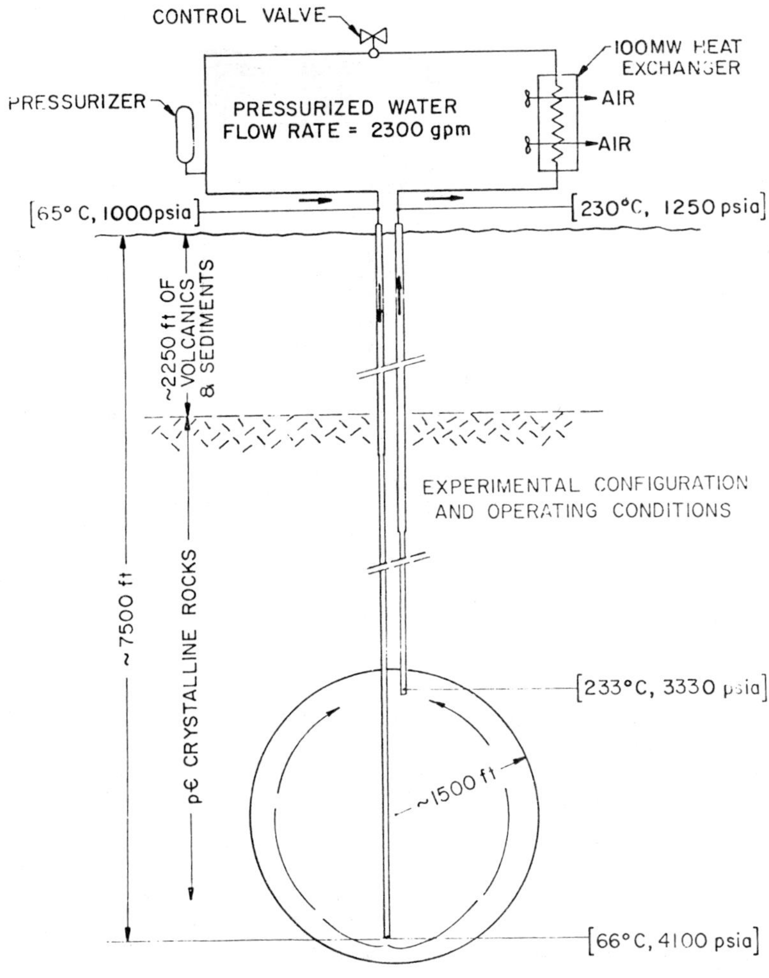

Figure 46. Schematic diagram of experimental work at Los Alamos.[19]

the multiple process utilization of a resource thereby increasing efficiencies and reducing costs.

Two processes have been developed based on the power cycles already devised: (1) an integrated system and (2) a divorced system, see Figure 49.

The integrated system involves wet steam brine being pumped through a well-head separator. The steam and brine then pass through a desalting plant where the steam is cleansed of hydrogen sulfide, carbon dioxide, and ammonia and water

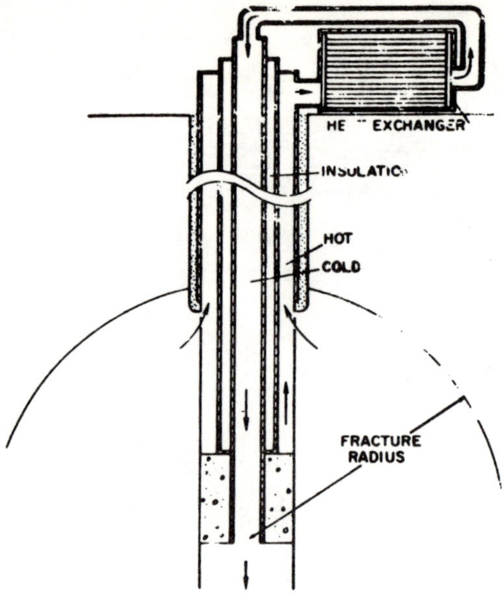

Figure 47. Concentric pipes system of withdrawing geothermal energy with hot water from a dry rock resource.[100]

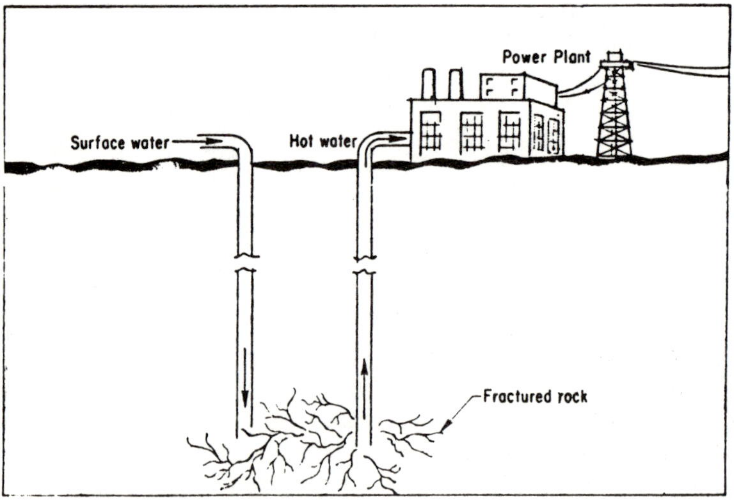

Figure 48. Simple artist's conception of a possible geothermal hot rock resource to be utilized for power generation.[14]

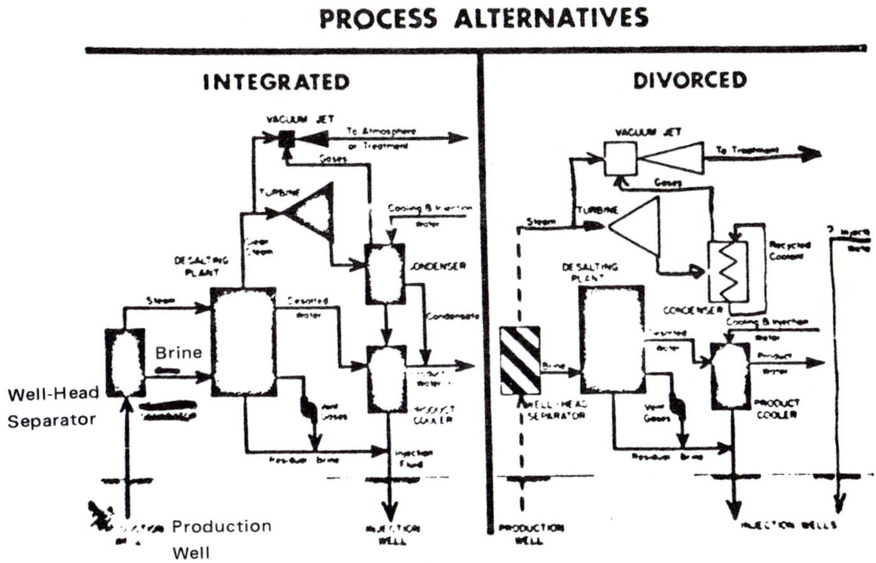

Figure 49. Integrated and divorced processes producing electricity and water from one power plant.[30]

separted from the brine. Clean steam is used to drive the turbine for power production and therefore conventional materials of construction can reduce costs. Condensing steam (from the turbine) is combined with the desalted water in the product cooler to effect pure water for a variety of uses including agriculture, domestic, commercial and industrial. The residual brine is combined with make-up cooling water and re-injected into the ground.

The divorced system is quite similar to the integrated system as far as processes go except that the individual electrical and water production facilities are separate.[30, 98, 101]

A more elaborate system of geothermal resource utilization is shown in Figure 50. The cycle shown is a typical flashed steam process power plant, but with the separated brine water and exhaust steam treated to deliver minerals, pure water, and excess heat which can be used for space heating, etc.

Figure 51 is a process similar to Figure 50 except incorporated into it is the hot water-secondary fluid power cycle. Both systems are the ultimate design concept of a geothermal power plant, as even escaping gases such as CO_2 can be utilized. These

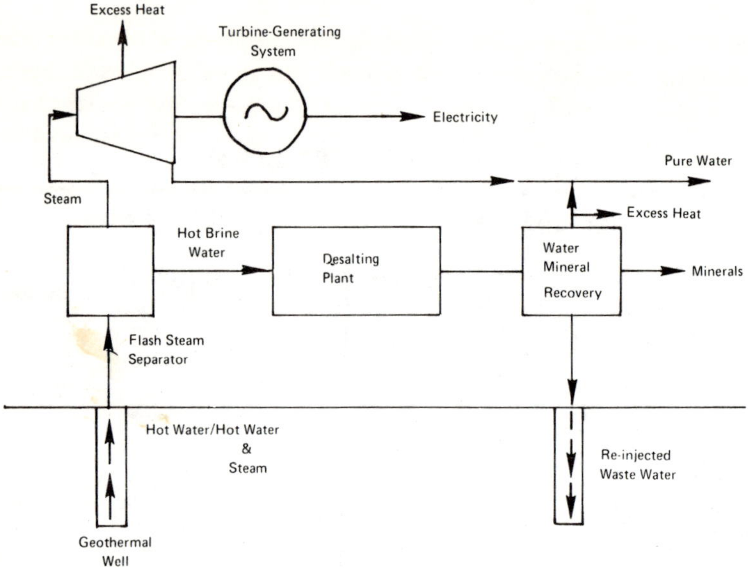

Figure 50. Geothermal power, water, mineral, heat plant using flashing steam from hot water type cycle.

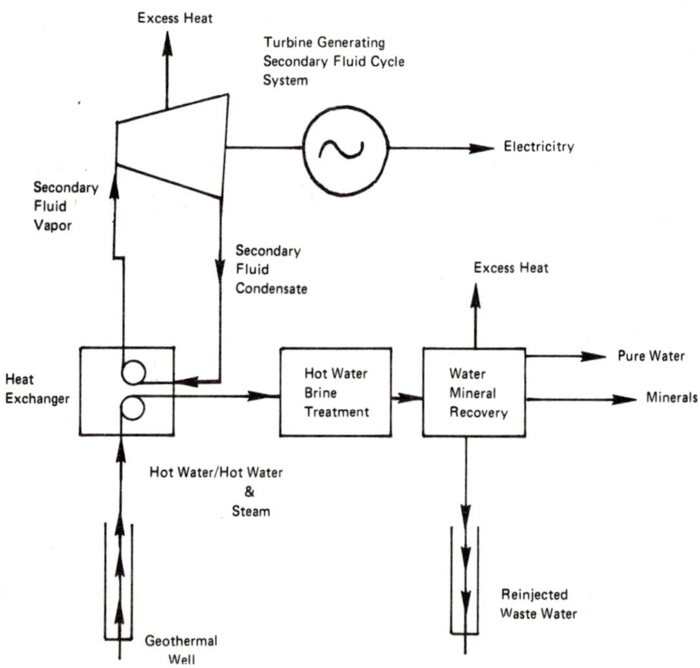

Figure 51. Geothermal power, water, mineral, heat plant using hot water/secondary fluid cycle.

systems will produce power, water, minerals, and chemicals, and heat for other uses in a practically pollution-free closed system. And as we examine further applications and usages of the geothermal resource, the reader will realize that such a system is completely viable technically and economically. The need is being created.

CHAPTER 6

GEOTHERMAL ENERGY SYSTEMS AND APPLICATIONS

Blessed of the Lord be his land. . ., for the precious things of the earth and fullness thereof: . . .for (the people) shall suck of the abundance of the seas, and the treasures hid in the sand. . . DEUTERONOMY 33:13, 16, 19

Geothermal energy is a highly versatile entity whose uses through direct and indirect means and their corresponding byproducts are countless. In fact, it will soon appear evident that the applications of a geothermal resource are bounded in many respects by the conceptual limitations of man.

The possibilities for the future include entire communities developing at a geothermal resource site. This community will utilize fresh water for domestic, commercial and industrial needs, the heat for heating and cooling, steam for electricity production and a variety of geothermal products and byproducts for agricultural and industrial applications. These communities could become industrial as well as agricultural centers through the usage of geothermal waters.

Fresh Water

Water as available for drinking is estimated at less than 2 percent of the earth's supply. The oceans, the atmosphere, rocks or rock formations (and polluted resources) contain the remaining 98 percent of the earth's water supplies. This supply remains constant as the world's population increases, however, consumption increases in proportion with the population growth and also with standard of living advancements. 400 billion gallons of water are used each day in the United States with a growth rate of 25,000 gallons/minute.[103] Thus, we have a fixed resource that is continually being drawn from and at increasing rates to effect water shortages across the globe. On the other hand, there are many regions as yet desolate which could support life and agriculture upon the addition of fresh and viable waters. Many of these regions are adjacent to geothermal resources and their ultimate development could have a severe effect on world population migrations and food supplies.

From the standpoint of water shortages; desalination, recycling and/or transportation over long distances of water has proved uneconomical. All the systems recognized to date consume enormous amounts of energy and most of these are run at the expense of fossil fuels. Thus, our problems are compounded since fossil fuels are in short supply which would be required to alleviate water shortages. Potable water parameters are not only set by medicinal and aesthetic requirements, but also by economics. Water is expected to be highly inexpensive, therefore, any purification process must not magnify water costs greatly.

Desalination, or the upgrading of saline waters to potable standards, is dependent on the energy cost for the conversion process. Distillation of the waters is a process that is the conversion process best developed from a technological stand, but is relatively inefficient as large amounts of energy must be used to boil large amounts of saline water. Obviously, nuclear energy will be utilized to power desalting plants as we enter a new era of expansive and disseminated energy source.[102, 103]

As already mentioned, many geothermal resources contain vast reservoirs of hot water and steam and some of these are producing electricity and fresh water as a byproduct. The geothermal resource satisfies the two main criteria for alleviating water shortages:

1) An energy source for distillation,
2) An ample supply of water.

The problems associated with the geothermal waters is their variation in characteristics. Water temperatures may vary between a couple of hundred degrees Fahrenheit to over 700°F. Salinity concentrations may be as low as 1000 ppm, but as high as 300,000 ppm. Both of these properties are dependent on well depth and at shallower depths the lower numbers for the above characteristics can be expected. Desalting operations are economically feasible if the saline waters are available at the surface with temperatures about 250°F. A typical well analysis is given in Table 24 for both the geothermal waters and steam.

Table 24. Typical Well Analysis, Casa Diablo.[104]

	Endogenous No. 4 (Water)	Endogenous No. 4 (Steam)
SiO_2	200	.8
Ca	4	–
Mg	–	–
Na	308	5
K	32	–
Li	.3	–
Fe	–	–
Al	–	–
B	11	.3
Cl	227	5
SO_4	96	2
H_2S	14	11

F	20	—
NH_3	.1	.5
CO_2	180	205
As	.2	—
pH	6.5	4.9

One geothermal desalination process and the most common incorporates a cycle for the joint production of electricity and water. Figure 52 shows of a typical desalting cycle being planned at Imperial Valley. A cyclone separator is in line to separate steam, both natural and that which has flashed from hot saline waters. The noncondensable gas such as hydrogen sulfide, carbon dioxide, ammonia, and hydrocarbons must be removed from the steam (see Table 24). A pollution problem which may be solved by reinjection of these gases with residuals from the water desalting process into the ground.

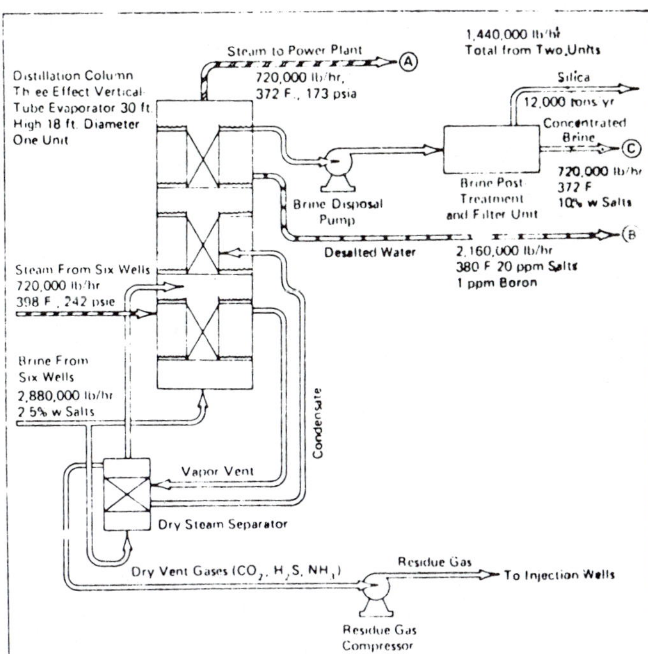

Figure 52. Typical schematic of a desalting cycle.[7]

The small scale project for Imperial Valley would effect the development of six 70 Mw plants and about 100,000 acre-ft of fresh water annually from 72 wells. A large-scale plan for this regions geothermal resource includes 1,800 production wells o feed 150 plants and 2400 reinjection wells, to provide 10,500 Mw of electrical power and 2.5 million acre-ft annually of fresh water.[7, 102]

Another fresh water geothermal process is shown in Figure 53. Hot well steam and saline waters and desert air are brought together in the evaporation chamber. The air becomes saturated with water vapor at 100°C. Blowers transport the air through the system and the hot saline waters are sprayed into the air stream. As the heat and moisture are transferred to the air, residual wastes collect on the bottom of the chamber and are drained. Depending on quantities of rejected residuals, the possibility exists for reinjection into the ground. The saturated air stream then travels through a thermally insulated pipe network to a cooling medium such as a river, lake or reservoir for condensation. Diffuses allow passage of the saturated air into the cooling water which removes the water vapor from the air as a result of the temperature differential. Figure 54 is a graph relating geothermal fresh water production to various volumes of saturated air at 100°C. Proposals for this system include the utilization of the Colorado River for eastern California geothermal condensation and the Feather River for western and northern California fresh water condensation sites.[104]

The advantages of the former system over this system is that by producing electricity and water and then using only 25 to 30 percent of this power production for internal needs the system becomes self-sufficient. The saturated air process requires electricity for blowers and pumps.

District Heating-Domestic, Commercial, Industry

We have mentioned heating from geothermal activity in previous chapters and have noted some places where its utilization is occuring. Hungary, Japan, New Zealand, and Russia have various entities heated with geothermal energy and encouraging plans for future expansion of the resources, however, none have a more widespread and diverse geothermal program as Iceland. Thermal waters in the temperatur range of 40–100°C can and are used to heat houses, commercial and industrial buildings. The thermal waters are transported many miles across Iceland to supply heat to over 80,000 people or 40% of Iceland's population, and nearly 90% of Iceland's capital city, Reykjavik is geothemally heated. Figure 55 shows the total production of geothermal energy in Iceland and its various applications.[105, 106]

Waters of a geothermal nature, although in many cases high in mineral content, can have limited use for hot water supply to showers and baths, washing, and depending on its quality cooling and cleaning. Househeating in the United States from hot water wells has been limited to Boise, Idaho and Klamath Falls, Oregon;[7] and while costs are impractical to compare during these times, houses

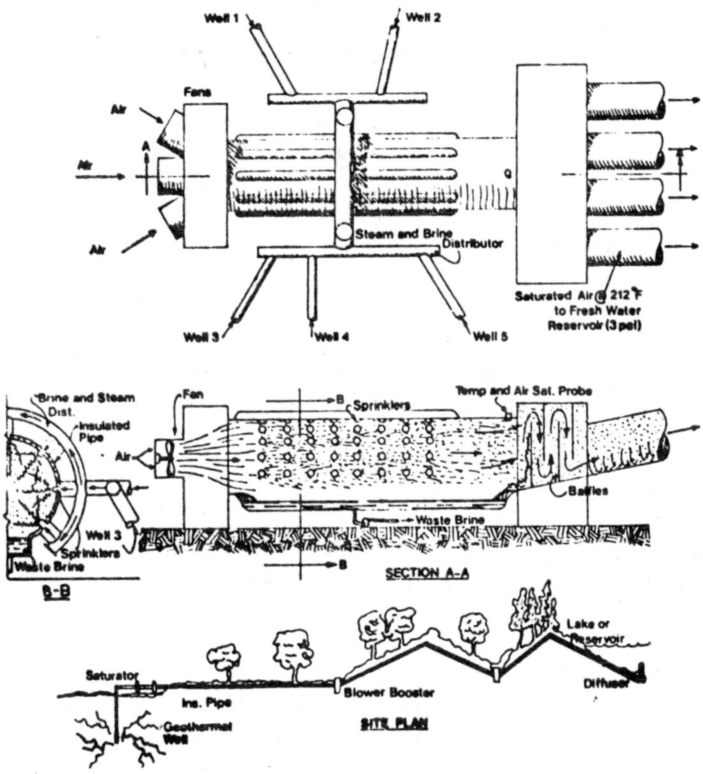

Figure 53. Geothermal-saturated air fresh water system.[104]

in Hungary heated in this manner have reduced costs by 1/4 (but this was before our current crises), more realistic numbers would show geothermal heat to be 1/10 to 1/20 the cost of conventional fuels.[29] More importantly, the amount of conventional fuels saved as a result of geothermal space heating and hot water supply will be considered.

The primary feature differences of thermal waters for heating is that these waters are used only once and passed through the system. These thermal waters have a constant temperature which is only dependent on well temperature. See Figure 56 for a typical house and building heating system. Geothermal water is pumped to a storage tank which then supplies the various buildings hooked into the system. The cool water is then reinjected into the ground as waste. Conventional heating systems recycle their heating mediums, (usually water or steam) whose temperatures can be varied depending on seasonal and daily temperature differences.

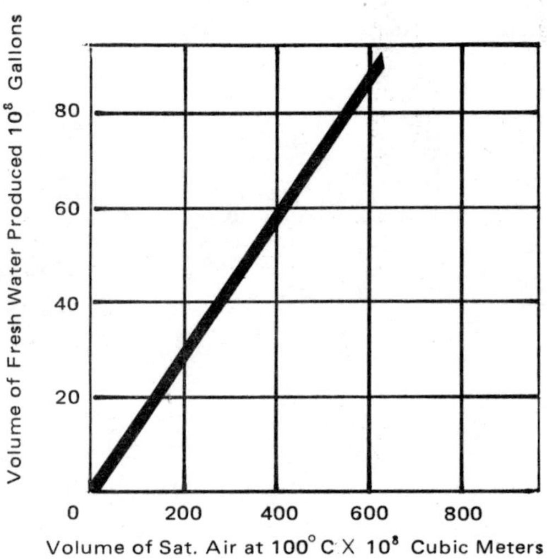

Figure 54. Geothermal fresh water production from saturated air.[104]

The total amount of conventional fuels saved by the use of geothermal heating waters is dependent on the type of application and kind of fuel customer. Ngeoth is the ratio of the actual amount of heat obtained by an actual customer to the amount of heat obtained by a standard customer. A comparison of efficiencies shows:

district heating	Ngeoth 0.05–0.34
ventilation	Ngeoth 0.15–0.45
hot-water supply	Ngeoth 0.76–0.92

Thus, we note hot water supplies are the most efficient use of geothermal waters and district heating can have rather low efficiencies, however, the proper application and combination of geothermal waters can increase efficiencies and thus save more conventional fuels, i.e. depending on the application fuel savings for a particular system may vary from 1200 tons of fuel per year to 4500 tons per year for a hot water supply system.[107]

Figure 57 shows the schematic of the Reykir fields, Iceland district heating system. The water is supplied to the houses for heating and hot tap water at about 80°C. The early systems used single pipe connections and the used water was discharged to the sewer system, however, newer systems using hotter waters have utilized two pipe house connections and cooler used waters are then mixed with

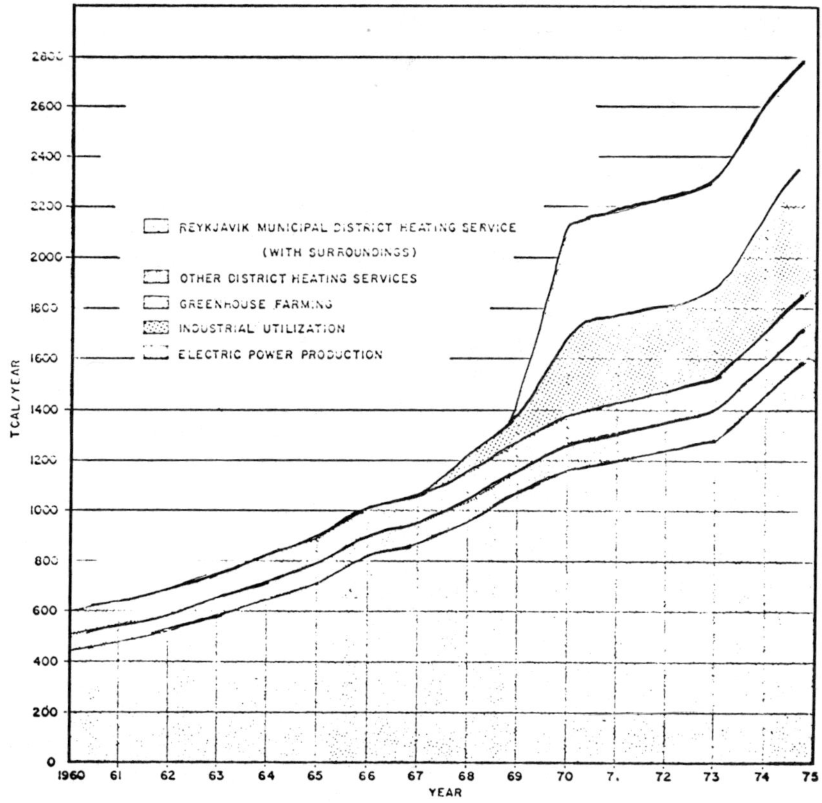

Figure 55. Iceland geothermal energy production.[106]

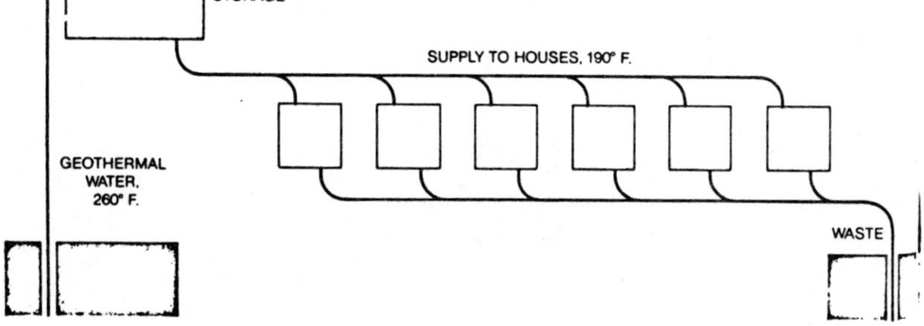

Figure 56. Typical geothermal heating supply to houses and other buildings.

hotter waters to obtain the desired heat level. Figure 58 shows a typical house connection up to a district geothermal heating system.

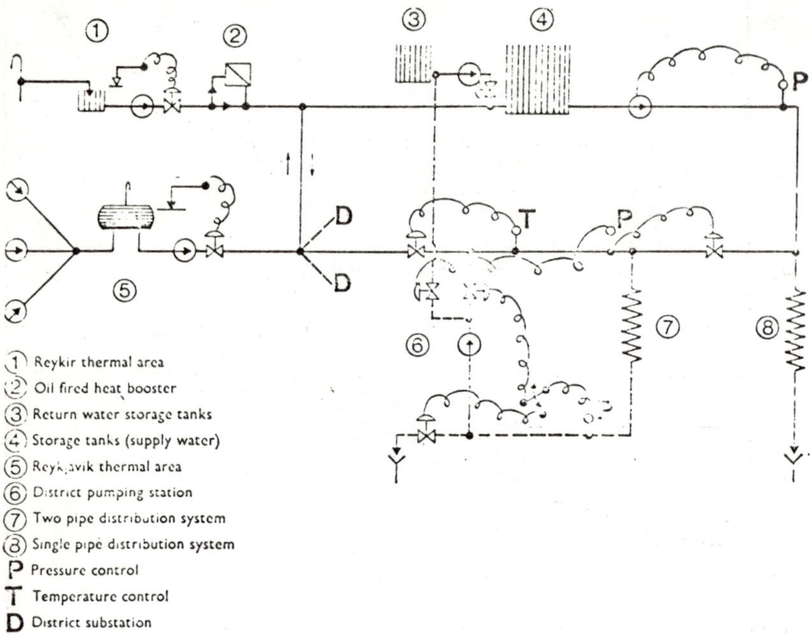

Figure 57. Schematic diagram of district heating system, Iceland.[105]

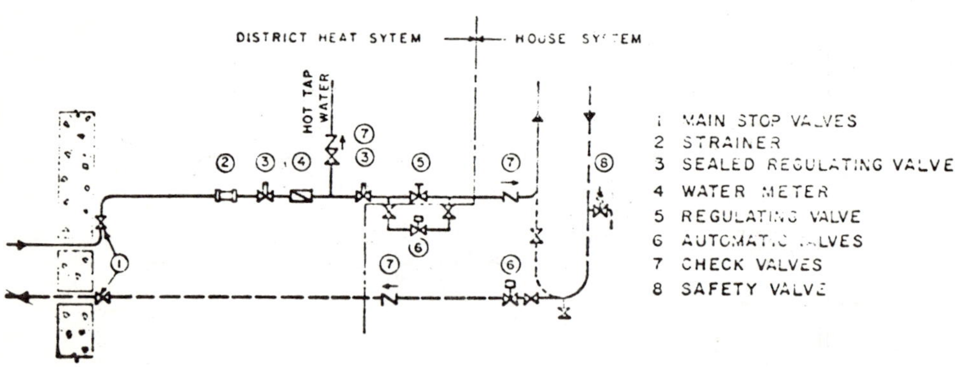

G

Figure 58. Geothermal heating house connection.[105]

Agricultural Uses and Systems

Three geothermal well constituents heat, water, and CO_2 content are combined for the complete utilization of the geothermal resource for agricultural purposes. Previous studies [29] have shown that greenhouses built with environments artificially controlled have several advantages in the production of crops including:

1) large crop yield increases;
2) crop production can continue year round;
3) agriculture workers are employed year round;
4) predators, pests, insects and frosts are controlled and can be eliminated.

Furthermore, CO_2 addition to a greenhouse atmosphere can result in increased crop yields by as much as 50 to 80 percent. By providing greenhouses with a continuous supply of heat, water, and CO_2 (refer to Table 24 and note the high CO_2 concentration) the geothermal resource becomes an invaluable agricultural tool

In Russia it was noted that 40 to 50 kg of coal is necessary to produce 1 kg of tomatoes or cucumbers in the winter in a heated greenhouse. The Soviet Union has expanded on the indoor agriculture technique and in one area thermal waters with temperatures of 60–65°C and quantities of 2000 m^3 per hour have resulted in greenhouses yielding 300–400 tons of vegetables annually. Many other regions of Russia have utilized this technology, with plans for the future for production of vegetables to reach thousnads of tons as a direct result of the geothermal resource.[40]

Many countries use thermal waters for agriculture including Japan, United States, Iceland, New Zealand, and Hungary. Japan's cultivating facilities are in many instances run by individual farm families with relatively small operations of 100 to 1000 meters.[2] Their produce includes melons, tomatoes, cucumbers, papayas, and other fruits and vegetables. Flowers include chrysanthemums, lillies, and other ornamental plants. Iceland has been exploiting the geothermal resource since the early 1920's. It has developed facilities of cultivating greenhouses that cover an area over 110,000 square meters. Two-thirds of this area results in produce and one-third flowers.

Geothermal technology is being extended as Iceland has been investigating the utilization of geothermal waters with temperatures as low as 30°C. The design of a low temperature pilot greenhouse for geothermal waters is being examined, and the final result will include the combining of space heating and soil heating technologies to determine the best working conditions and the construction of low cost greenhouses.[108] Figure 59 shows the schematic for such a greenhouse and its eventual hookup into the district heating system or possibly into one well or well field system. The fantastic benefit of the above system is that normally discarded waters from homes and buildings, which the district heating system discards because of their low heating value, can be used in the growing of crops.

Thus, hot water is first used to heat homes and then used to eventually feed their occupants.

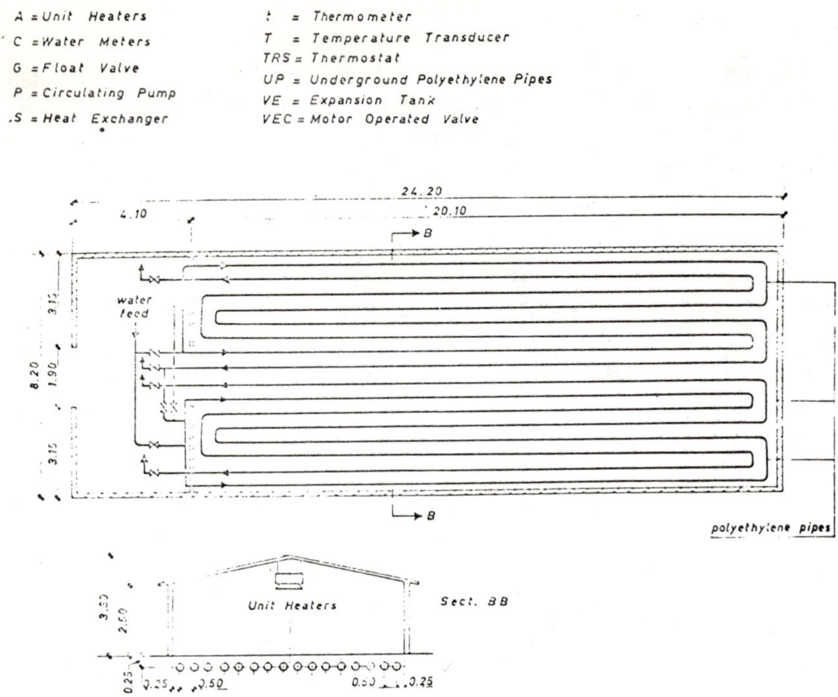

Figure 59. Schematic of an Iceland greenhouse for the total utilization of the geothermal resource.[108]

The criteria for the eventual constriction of a low cost-low temperature greenhouse are based on investigations of the following:

1. The dependency of greenhouse efficiency on soil heating;
2. The heat exchange between the soil and the polyethylene heating pipes;
3. The effect of water temperature variations and its relationship to soil temperatures on greenhouse efficiencies;
4. The quantities of heat directed to the soil for the greenhouse vegetation;
5. The minimum soil temperature that effects the best greenhouse efficiency;
6. The minimum water temperature and flow rate for maximum greenhouse efficiency.

Large volumes of geothermal water, obtained in manners previously discussed in this chapter, are used for the irrigation of lands and direct crop growing. The exploitation of a geothermal resource for irrigation purposes can result in the formation of new and enormous fertile basins for agriculture, heretofore laid barren due to lack of adequate rainfall. Waters of low temperatures, extracted from a home heating or electrical generation process or originating in that form; can be pumped directly onto a field and is cooled only from open flow. Sprinkler irrigation has become popular in southern Idaho[109] since the waters cool more rapidly as the spray settles through the atmosphere. Water of this nature is also used for animal breeding; including cattle and although geothermal technology is not on a par with food processing technology, geothermal energy can be applied to all facets of food generation.

Through watering, heating, and/or chemical additives for plant growth (both gases and fertilizers) the geothermal resource can support food production, processing, preserving and storing (see Cooling with Geothermal Energy).

Cooling with geothermal energy is another new technology which will be developed as interest in this energy source grows. One of the major cooling installations is in New Zealand.[110] The Tourist Hotel in Rotorua, New Zealand is fully heated, air conditioned and supplied with hot water from a single geothermal resource. The heating requirement was 2 million Btu/hour and hot water requirement, 2 million Btu/hour. The cooling system absorption unit required 2.3 million Btu/hour to effectively air condition the hotel. Figure 60 shows the piping schematic for the hotel air conditioning system. The technology developed from this hotel can be further applied on a large scale to produce a district heating *and cooling system.*

An industrial use of this geothermal process is refrigeration for food preservation and storage. Figure 61 shows the basic flow diagram for a geothermal refrigeration system. This system is effective to minus 60°C. Whether the system shown is capable of being modified such that individual refrigerators and freezers can operate on its principle depends on an expansion of technology in this area.

Industrial Applications

The ultimate exploitation of geothermal resources will occur when the obvious industrial and agricultural uses are finally realized. Many industries use large amounts of steam in their production processes and could eventually adapt these processes to geothermal steam. Table 25 shows numerous products and processes and their corresponding steam requirements. Steam requirements shown are per pound of product and for the unit dollar value of the product. Product values are shown, but these figures are somewhat outdated and do not reflect the extreme inflationary period of recent days. Obviously the huge increases in fossil fuel prices have increased steam costs at least three-fold and therefore less expensive steam sources have become desirable. However, the attractiveness of geothermal steam as well as other geothermal energy sources to industry is dependent on an improving technology capable of adapting many new processes, and the publicizing of currently available systems which utilize geothermal energy. Some industries using a geothermal process will now be discussed.

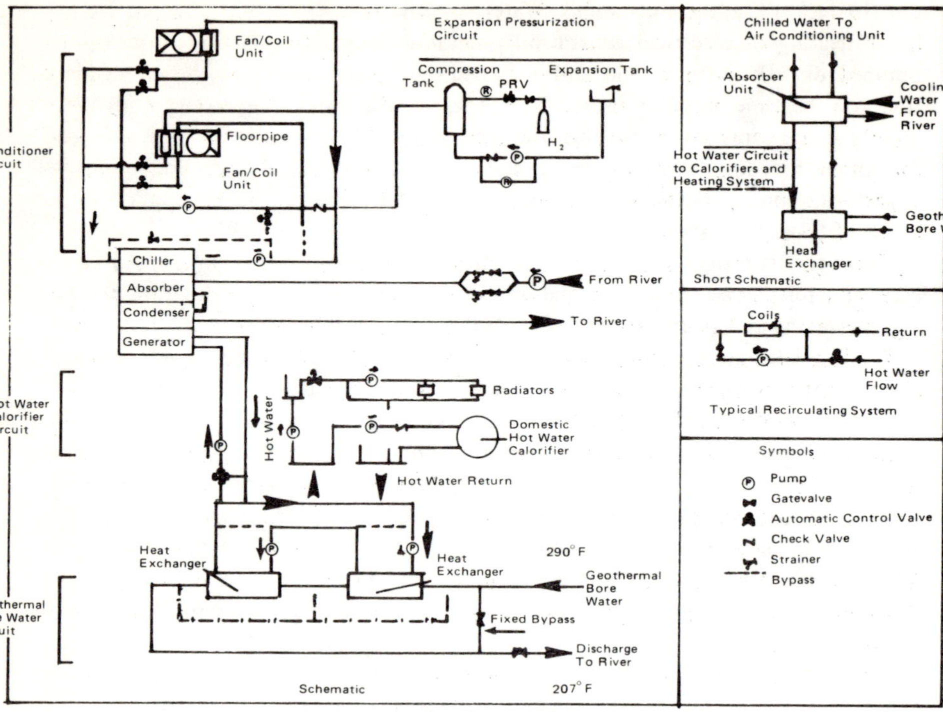

Figure 60. Geothermal heating and cooling system piping schematic, Rotorua Hotel.[110]

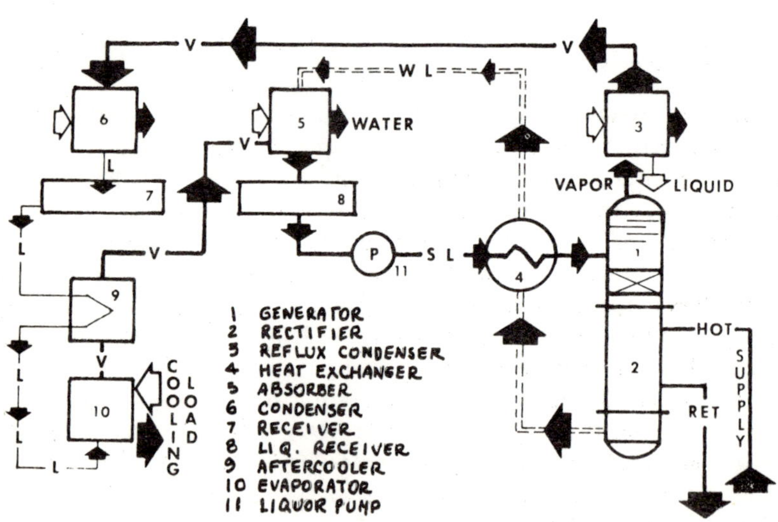

Figure 61. Geothermal refrigeration system flow diagram.[10]

Table 25. Steam Consumption for Various Products and Processes.[111]

Product and process	Steam requirements lb steam/lb	Product value cents/lb	Steam per unit product value lb steam/ $value
Heavy water by hydrogen sulphide process	10,000	3,000	333
Ascorbic acid	250	250	100
Viscose rayon	70[1]	75	93
Lactose	40	14	286
Acetic acid from wood via Suida process	35	10	350
Ethyl alcohol from sulphite liquor	22	7	314
Ethyl alcohol from wood waste	19	7	271
Ethylene glycol via chlorohydrin	13	13	100
Casein	13	56	23
Ethylene oxide	11	15	73
Basic Mg carbonate	9	11	82
35% hydrogen peroxide	9	18	50
85% hydrogen peroxide from 35% H_2O_2	4¾	—	—
Solid caustic soda via diaphragm cells	8	3	266
Acetic acid from wood via solvent extraction	7½	10	75
Alumina via Bayers process	7[2]	3	234
Ethyl alcohol from molasses	7	7	100
Beet sugar	5¾	10	58
Sodium chlorate	5½	9	61
Kraft pulp	4 1/5	6	70
Dissolving pulp	4 1/5	—	—
Sulphite pulp	3½	6	58
Aluminium sulphate	3½	2	175
Synthetic ethyl alcohol	3	7	43
Calcium hypochloride, high test	3⅓	3	111
Acetic acid from wood via Othmer process	2¾	10	28
Ammonium chloride	2¾	6	46
Boric acid	2¼	5	45
Soda ash via Solvay process	2	1½	133
Cotton seed oil	2	10	20
Natural sodium sulphate	1 4/5	1½	120
Cane sugar refining	1 2/3	10	17
Ammonium nitrate	1½	3½	43
Ammonium sulphate	1/6	1½	11
Fresh water from sea water by distillation	1/12	1/60	500

Pulp and Paper

The paper and pulp industry uses casts amounts of water both in the liquid and vapor states. Initially it is used in the digesting of the wood, then in separation and recovery of byproducts and finally in washing the end product. A geothermal resource would seem naturally oriented to the paper and pulp industry, a fact that has not gone unnoticed in New Zealand.

The Tasman Pulp and Paper Company of New Zealand has been applying geothermal steam to its paper processes for a number of years. Its mills produce newspring, kraft pulp and sawn timber. Presently, geothermal steam contributes over 50 percent of the requirements of the milling complex. 400,000 lb/hr of geothermal steam along with 350,000 lb/hr of steam produced from conventional methods generate electricity and process the wood into paper.

Characteristics of the geothermal steam at the mill of pressures of 100 psig and 200 psig and steam and water mixtures with enthalpies between 430 and 624 Btu/lb are normal. The steam and water are separated with cycline separators and steam is also flashed from the hot water. The main feature of this plant which must be stressed is the combined contributions of the geothermal resource and the conventional resource to achieve the final results. This fact must be further emphasized, that geothermal energy is not the "cure all" for the world's energy problems, but by allowing it to contribute its fair share through increased technology and utilization we can alleviate many of the shortages. Figure 61 shows the pulp and paper plant and the geothermal field contributing steam in a piping network along with steam produced in boilers to the plant.[112]

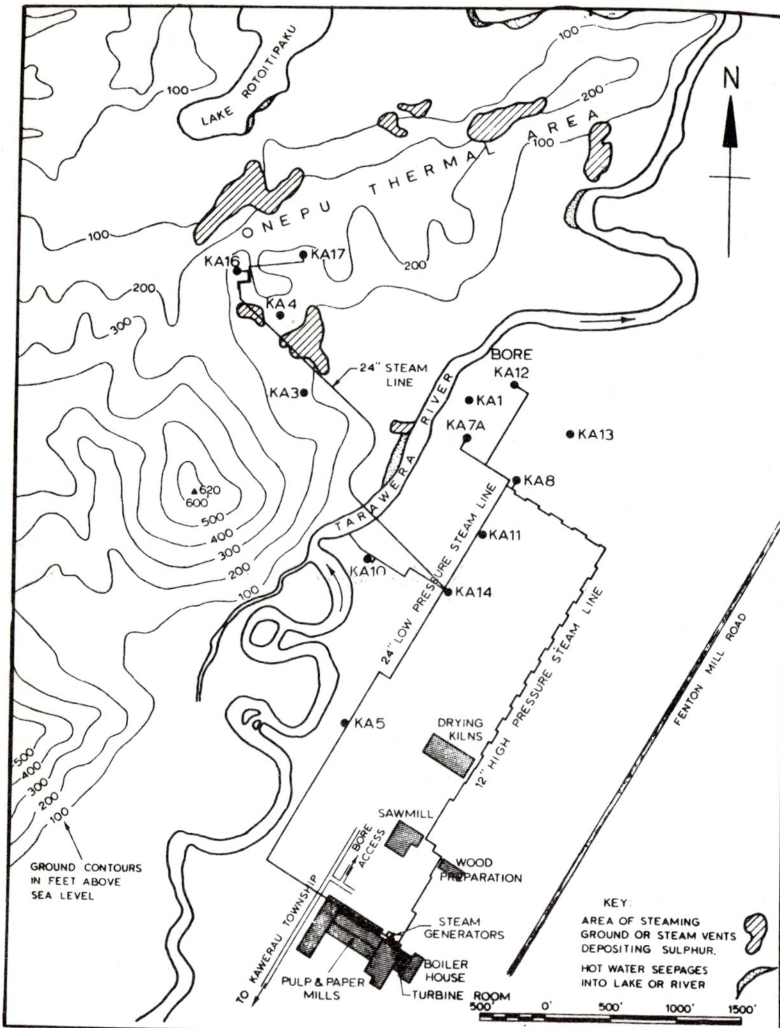

Figure 61A. The Tosman Pulp and Paper Company geothermal steam plant complex at Kawesau, New Zealand.[112]

Other Systems, chemicals and minerals

Also requiring large amounts of steam is the sugar processing plant. The application of geothermal waters to the sugar process can result in considerable savings over conventional fuels as well as water conservation.

Diatomaceous earth in Northern Ireland is dredged from the bottom of Lake Myvatn. Natural steam heats and dries the diatomaceous earth until a final product has evolved. Figure 62 shows a schematic of the diatomite plant.[113] Thus we note the availability of geothermal energy to mining and mineral extractions techniques. These waters can also be used in the final refining of the minerals.

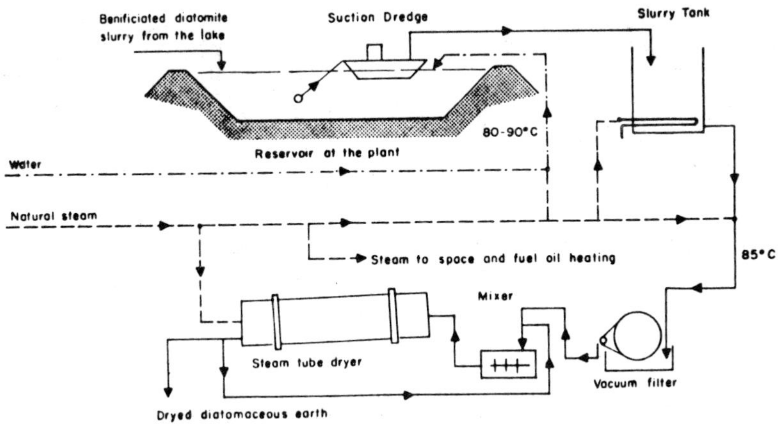

Figure 62. Schematic of a diatomite plant in Iceland which utilizes geothermal steam.[113]

As a result of the high water temperatures many geothermal fluids contain high concentrations of valuable minerals and chemicals. Aside from the production of salt which we more or less considered as a by-product of the fresh water process, the constituents of a geothermal steam are many and some are cited in Table 26.

Table 26. Geothermal Resource Constituents.

Calcium	Iron	Carbon dioxide
Ammonia	Arsenic	Bromine
Sodium	Mercury	Iodine
Potassium	Hydrogen Sulfide	Strontium
Boron	Silicon	Lithium
Magnesium	Nitrogen	Naphtenic acids
	Other trace elements	

A process has been devised for the utilization of geothermal waters in the production of magnesium chloride:

$$Mg(OH)_2 + NaCl + CO_2 \rightarrow MgCl_2 + Na_2CO_3 + H_2O$$

Some purification of the waters must take place before they would pass through the system, however, these problems are considered minor to the ultimate development of the project. Figure 63 is a flow diagram of the magnesium plant.

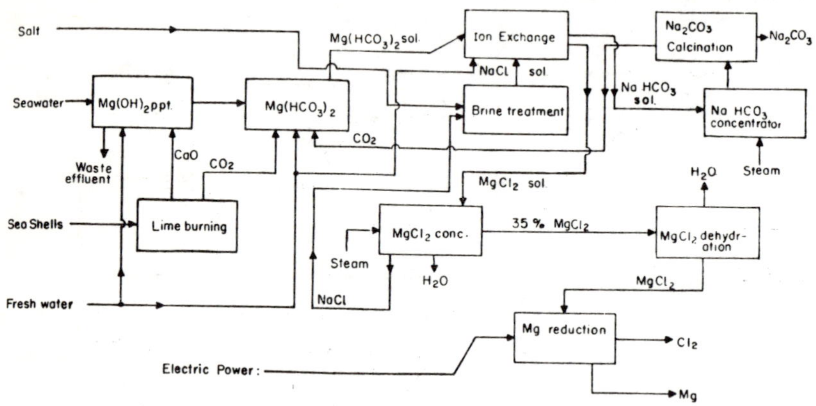

Figure 63. Flow diagram of a possible magnesium plant.[115]

The possible complete utilization of a geothermal brine (geothermal waters with high dissolved solids content) is shown in Figure 64. Shown in Figure 64 are mineral extractions, metal extractions, chemical production, and chemical processes.

Health and Spa Resorts

In the first chapter, we explored some of the possibilities of geothermal energy and found that the oldest and most widespread utilization of geothermal energy is for its balneological and recreational purposes. Thousands upon thousand, in every country, with an availability of geothermal energy, of resorts have been built on a hot spring or boiling water system.

In addition, water bottling plants sell this high mineral content across the globe. Thus, a profitable business evolved, from this resource, however, other explorations have not proved as fruitful, but the technology for this enterprise has been developed over thousands of years – since ancient Rome and ancient Japan – the technology for current geothermal uses has not had a couple millinium of refine its techniques. Therefore, the eventual mastery of the resource is apparent but is directly proportional to need, availability, costs, and the desire to improve our current status.

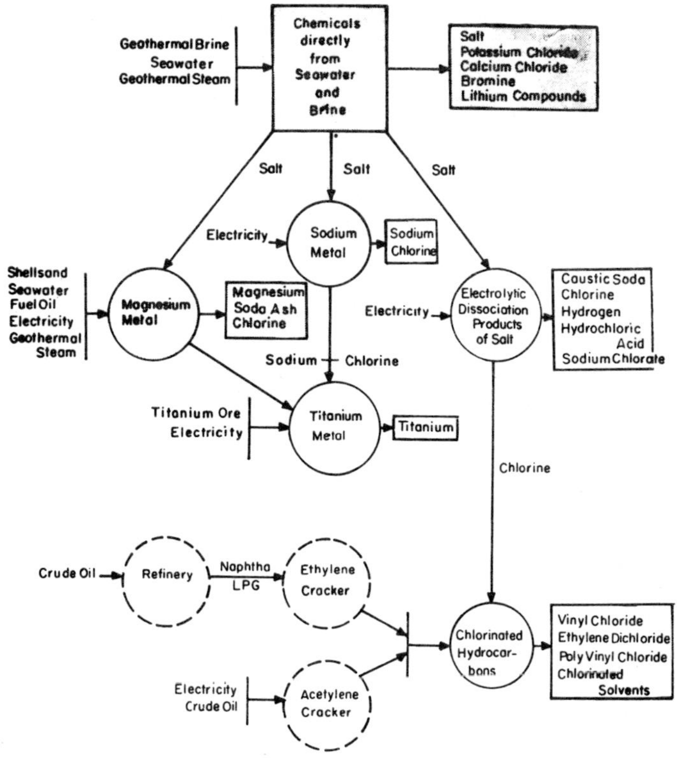

Figure 64. Chemical and mineral development from geothermal fluids.[115]

CHAPTER 7

GEOTHERMAL POWER PLANT DESIGN

He turneth the wilderness into a standing water, and dry ground into watersprings. And there He maketh the hungry to dwell, that they may prepare a city for habitation . . . PSALMS 107:35, 36

Endogenous Fluid

The initial phase of geothermal power plant design involves determining the steam availability, and this includes examining:

1. The location of the geothermal field and high density steam areas.
2. The quantities of steam available.
3. The fluid quality; whether it is a dry steam or wet steam or just plant hot water. If the endogenous fluid is of the latter two qualities, alternate designs including auxilliary equipment must then be considered.
4. Other characteristics; temperature, pressure, available heat content.
5. Impurities and non-condensable gases content of the steam. Determinations of the type of impurities and gases as well as the quantity.

A primary consideration to the power plant design, and we shall take a look at what the Italians have done and propose to do about it, is the geothermal field longevity. It is generally believed that the life of a geothermal source is only dependent on the continuous heating of the source by the earth (which for our purpose can be considered infinite). Also, as we have learned, the resource is a continuous cycle with cold surface waters travelling to the subsurface, being heated, and then reappearing at the surface. However, there are steam variations with each well and with time. The Geysers, Larderello and Waikaikai, have all experienced a pressure drop during their years of operation, necessitating expansion of the geothermal field. The power plant must be flexible enough to take in a variety of steam characteristics and impurities and allow for changes in steam flow rates.

Thus, quantitative and qualitative preliminary studies have to be made before any geothermal power plant can be designed. The results of these studies provide the basis for plant design and individual studies of each field must be made. Therefore, the Geysers' region underwent extensive examination. It was discovered that the most economical use of the Geysers' field would include the use of condensing steam turbines at a steam inlet pressure of 100 psig and an exhaust pressure of 4 in. of Hg. The steam at the Geysers was found to be slightly superheated with a constant enthalpy of 1200 BTU/lb. As a result of these parameters, higher pressured steam increasing the available steam energy would decrease the steam flow rate and lower pressured steam increasing the steam flow rate would decrease the available steam energy, thus the optimum steam characteristics were defined.[8, 13, 31, 117]

Construction Materials

Construction materials of a geothermal power plant vary with respect to the system process. That is, initially the steam is superheated as it leaves the wells with non-condensable gases. Under these conditions the mixture is relatively non-corrosive. Therefore, the piping from the wells to the turbines and the turbines themselves can utilize a carbon steel alloy and reduce construction materials cost. However, in the event of wet steam oxidizing some H_2S into sulfuric acid, non-corrosive materials must be used. After the steam leaves the turbines it passes into the condensers and condenses. The non-condensable gases become highly concentrated, the hydrogen sulfide concentrated, the hydrogen sulfide oxidizes partially to become sulfuric acid, and the corrosiveness to the power plant system necessitates the use of corrosion resistant materials such as austenitic stainless steel, aluminum and epoxy-fibre glass. Contact with the atmosphere changes sulfided to sulfates and the less expensive construction materials cannot be used, i.e. carbon steel, cast iron, copper alloys, zinc, cadium, silver. A coating of coal tar epoxy is required for concrete.

Hydrogen sulfide in the steam in the presence of air effect problem conditions to electrical equipment; relay contacts, exposed copper wiring, switches and other motor control gear and exposed metal parts. This problem is so acute that protective alloy coatings and aluminum have to be utilized, and the control room pressurized and filtered to reduce corrosion effects.

The cooling towers are extremely susceptable to the above effects and are designed with corrosion resistant materials. Wood has been shown to deteriorate in the presence of a geothermal atmosphere. Other cooling tower fill material has proven less susceptable to deterioration - polystyrene, polypropylene, or PVC which is also fire resistant.

As more and more studies are performed, proper material selection will become cataloged for each specific geothermal field type. Moreover, new corrosion resistant materials, some precious and semi-precious, have become incorporated into the geothermal plant as improved material design techniques surface.[8, 13, 31, 93, 117]

Design Criteria

Certain criteria for a geothermal power plant have been developed and these include:

1. The determination of an adequate rating of the geothermal power plant unit, based on future needs and field potentials. The elimination of large numbers of unit installations and/or of construction of large piping networks is the result of the proper unit power rating selection.
2. The ability of the plant to be rapidly installed to reduce energy losses due to escaping steam and to capitalize on the initial exploratory and research expenditures as soon as possible.
3. The power plant must provide for operational simplicity which ultimately will enable it to be automated and/or remotely controlled.
4. Machinery and auxilliary equipment must be mobile to allow for easy transfer of a plant because of new steam sources or the reduction of pressures of an old site prove to be uneconomical.
5. To allow for varying steam temperature, pressure, impurities and noncondensable gases the turbine and other equipment must be highly flexible to obtain high efficiencies.[118]

The Italians have set the above requisites for their geothermal power plants and appear to be the forerunners in plant design research. They have studies geothermal power plants to develop more efficient and practical turbines, condensers, piping and other auxilliary equipment.

The typical geothermal power plant is of conventional design. Power cycles (see Chapter 5) are unique only due to the fact that the steam is dusty and mixed with other impurities requiring added separation devices. The steam turbines are typical power plant turbines made of standard low temperature, low pressure materials. 11–13% chrome steel turbine blades and nozzles are used with carbon steel turbine casings. Stainless steel inserts are provided in the casings to prevent erosion.

Direct contact condensers are used at geothermal power plants because the geothermal steam is not recycled to boilers. The condenser parts which come in contact with the corrosive condensate are made of stainless steel, the outer shell can be made of less expensive carbon steel. Two types of direct contact condensers are in use: 1) the barometric and 2) low level direct contact condensers.

The barometric condensers are located above the turbines and the steam enters the bottom of the condenser shell. Water from the cooling tower spills over the steam in the counter flow condenser. Noncondensable gases exit from the top of the condenser and the steam and cooling water mixture leave by way of a barometric leg, (a large and expensive duct), maintained at atmospheric pressure and are pumped to the cooling tower.

The low level direct contact condenser utilize perforated trays to effect condensing of the steam, which flows counter to the cooling water. The steam enters the center of the condenser, condensed on its sides and then pumped to the cooling tower. Depending on turbine size and orientation one of the two condenser designs becomes more economical.[13, 93, 94, 117]

Turbine design and power plant cycles are functions of the endogenous fluid. Where condensing steam units are for steam with leas than 10% noncondensable gases, back pressure turbines are for steams with greater quantities of gases. Other turbine designs include variations of the above designs dependent on:

1. Temperature
2. Pressure
3. Steam Quality
4. Steam Impurities

Table 1 shows some Larderello area condensing turbines and some operating parameters. Table 27 shows some back pressure turbines and some of their operating parameters and characteristics. Note the magnitude of condensing turbines as compared to back-pressure turbines. While condensing units are necessary and more economical for most geothermal steam characteristics, back-pressure turbines have been able to utilize steam from wells heretofore thought uneconomical. Furthermore, the back-pressure turbines are transfereable from place to place, usually they are installed initially on a well site until steam characteristics change and a condensable unit is needed. It is then transferred to another well site. The impulse back-pressure turbines with their generators can be transferred completely. The reaction turbines are only partially transferrable due to excessive size and weight, the generator and other equipment must be transferred separately. Some typical operating characteristics of a back-pressure turbine are noted in Table 28.[95]

Table 27. Condensing Turbine Operating Parameters at the Larderello Field.[95]

N. of units	Output (kW)	Inlet pressure ata	Gas content %	Condenser vacuum inches of Hg	Location
2	1000	4	4	3″	Sasso Pisano Monterotondo
2000	1	4	4	3″	Larderello
3500	1	4	4	3″	Sasso Pisano
9000	2	4	4	4″	Larderello
12000	1	4	4	4″	Larderello
12500	2	4	4	3″	Sasso Pisano
14500	3	5	8	3″	Monterotondo Larderello - Lago

Operation

The design of a geothermal power plant is usually to allow for a minimum of

Table 28. Back Pressure Turbine Operating Parameters and Turbine Type.[95]

Number of units	Output kW	Inlet pressure ata	Type	Characteristics
5	3000	4	reaction	semitransferable
2	3500	4-8	impulse-reaction	variable-pressure semitransferable
1	900	4-8	impulse	transferable
1	3000	4	impulse	transferable

operating personnel. The power plant at the Geysers generally operates with very little maintenance. While the entire system is afforded continuous service, each individual unit is intermittently coverd since only 16 workers are required for 24 hr/day, 7 days/week maintenance.

Start-up of a well requires several hours of steam exhausting to the atmosphere to cleanse it of unwanted water and debris. Condensate must be drained from the piping network, and the piping itself must be warmed up. These preliminary procedures are required to prevent turbine fouling and damage. As a result of the dust mixed with steam, piping, valving, strainers as well as turbines and turbine accessories are inspected often. The turbine units are inspected twice as often (i.e. 3 years to 6 years) as fossil fueled plants and values may be examined numerous times daily. Lubricating of the system is most frequent. Rock particles rising in the steam and entering the turbine can cause blade pitting and wear. Replacement of certain parts becomes routine maintenance. Shutting down of a unit for short time periods results in the exhausting of steam to the atmosphere. In the event the shutdown is for relatively long time interval, two to three weeks, the wells are capped to preserve reservoir properties.[8, 13, 31, 93]

A typical geothermal power plant is shown in Figure 65. Note the placement of the barometric tube and the recirculting water system for condensation.

Steam Pipelines

The steam pipeline and/or hot water pipeline design is very critical to the geothermal resource. The well output characteristics, flow rate, pressure and temperature and correlated along with pipe flow research and design procedures to obtain the optimum piping design.

The initial calculations for piping consider the output available from the endogenous fluids. The total practical thermal yield, or the transformable thermal energy into electric energy is given by:

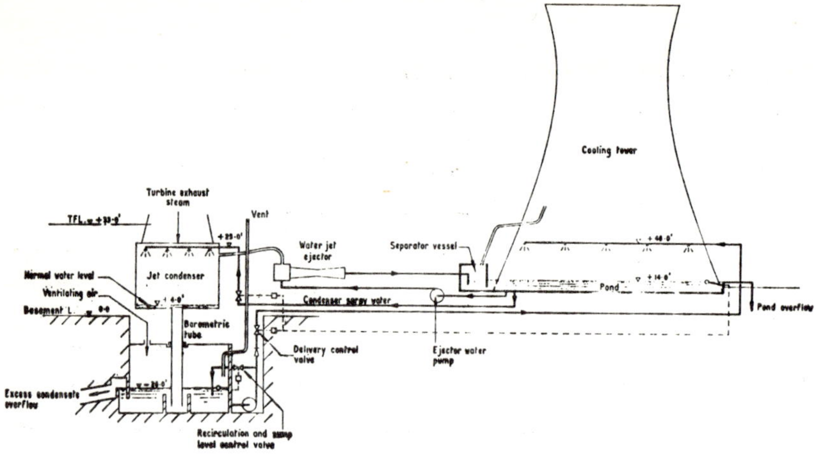

Figure 65. Power plant recirculating water system.[119]

$$R_{to} = \frac{G_n i_{fo}}{N_n}$$

where: R_{to} = total practical thermal yield
G_n = hourly delivery of fluid
i_{fo} = specific enthalpy of the fluid at the well
N_n = electric power obtainable at the well

The total thermal efficiency of the power plant is given by:

$$Ng = \frac{860}{C_s i_{fo}}$$

where: Ng = total thermal efficiency
c_s = total specific consumption of the power station.

Thus the thermal efficiencies at the well can be compared with the actual thermal efficiencies to determine the pipeline losses due to the transportation of the steam. The heat losses through the piping insulation are calculated:

$$Q = \frac{\pi(t_1 - t_2)}{1/2\lambda \ln d_1/d_o}$$

where: Q = heat losses per hour and square meter of thermal insulation (K cal/m^2h)
t_1 = temperature of endogenous fluid (°C)
t_2 = ambient temperature 9°C)
λ = insulating material internal conductivity (K cal/mh°C)
d_1 = thermal insulation outside diameter
d_o = thermal insulation inside diameter

From these equations[120] the most economical insualtion thickness can be determined. That is, the thickness obtained before increasing the insulation thickness costs cannot be justified by increased thermal and electrical output. However, standard pipeline diameters will eventually prove most economical at the expense of some heat loss.

The second consideration given in the pipeline design is endogenous fluid constituents. Chemicals such as chlorine and hydrogen sulfide in a wet atmosphere become highly corrosive. Other compounds form deposits which eventually block steam passage. Mechanical erosion of the piping is effected by dust and rock particles in the steam and by changes in flow. The latter phenomena can cause serious damage to the piping especially at the joints, elbows and tees. Special shock absorbing anchoring techniques have been developed to alleviate the possibility of damage due to sudden flow stops and starts. Figure 66 illustrates geothermal pipeline design. Pipelines must go over and under obstacles, and up and down gradients.

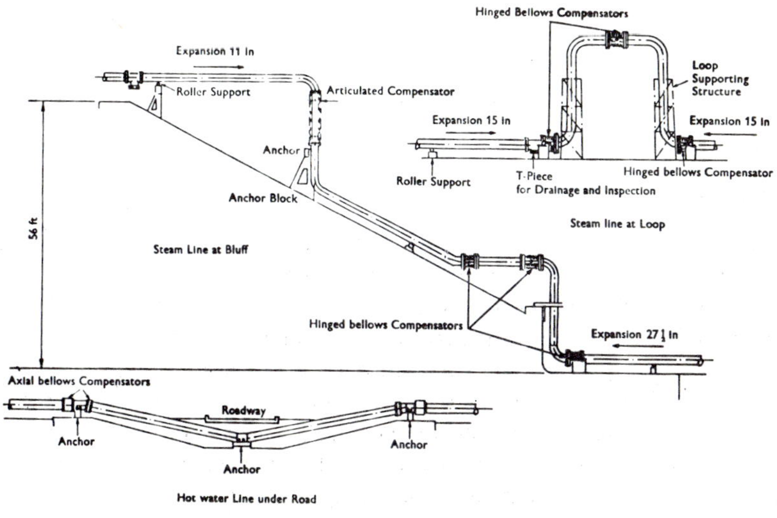

Figure 66. Geothermal piping design.[121]

Pipeline design criteria is dependent on land topography and distance of transportation. Endogenous fluids are transported in many instances short distances to the power generating stations from a few hunderd feet to thousands of feet to many miles as required in Iceland by its hot water district heating. No appreciable losses are effected by the relatively long distances the hot water of Iceland must travel. Pipeline technology is keeping pace with geothermal energy exploitation.

Figure 67 shows rather simply the piping configuration as the steam leaves a well. Figure 68 is a hot water transmission system. The diagram illustrates the complex piping necessary at a well site power station, but also that this piping must meet varying demands such as pressure and temperature differentials and a varying endogenous fluid.

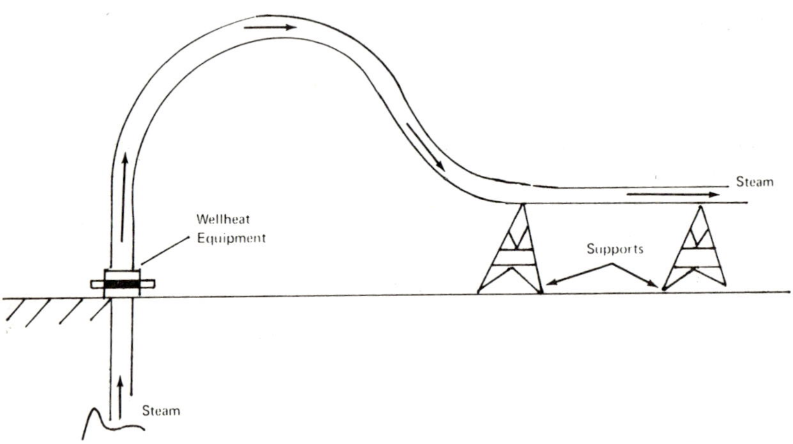

Figure 67. Typical well pipe configuration.

Economics

The economics of a geothermal power plant are comparable to other electric generating stations. Fixed charges of a geothermal power plant are about the same as conventional fossil fuel plants and less than half of a nuclear or hydropower plant. Other charges such as maintenance, operation, and steam or fuel were nearly on a par with other plants, some being greater and some more, depending on the methods of calculation. However, any comparison of costs between energy generation systems is not feasible because of the nearly constant upward spiral of conventional fuel costs. Pollution problems and the inflated crude oil prices have contributed to the large fluctuations in electricity costs.

A twelve year average of geothermal costs were taken between 1961 to 1972 inclusive at the Geysers power plant and are shown by Table 30. These costs will

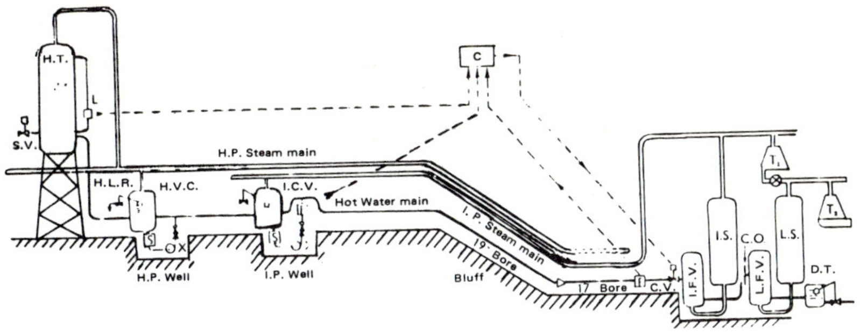

KEY:

H.C.V. H.P. collection vessel
H.C.V. I.P. collection vessel
H.T. Head tank
S.V. Motor operated spill valve
L. Level detector
F. Flowmeter
H.L.R. High level relief
S. Stramer
X. Extraction pump
C. Computer
== Steam connections
__Hot water connections
---- Electrical connections

C.V. Motor operated control valve
L.F.V. I.P. flash vessel
L.F.V. Low pressure flash vessel
L.S. I.P. Scrubber
L.S. L.P. Scrubber
D.T. Drain tank
T_1 I.P. turbine
T_2 L.P. turbine
C.O. Control orifice
H.P. High pressure
I.P. Intermediate pressure
L.P. Low pressure

Figure 68. Diagrammatic hot water transmission.[121]

Table 29. Typical Back Pressure Turbine Operating Characteristics.[95]

— Turbine speed	3000 rpm
— Turbine output	15 MW
— Inlet pressure range	5-8-11 ata
— Inlet temperature range	180 ÷ 250 °C
— Inlet gas content	0 ÷ 30% weight
— Exhaust pressure	atmospheric
— Turbine type	first stage impulse type other stages-reaction type

not vary greatly, even in inflationary times, because the only variable cost is dependent on steam input. Other costs are constant except for possible raises to personnel of which there are few. Two basic parameters will govern geothermal cost consistency.

Table 30. The Geysers Operating Costs Averaged Over 12 Years.[31]

Costs	Power Cost, in mills per net Kw-hr.
Capacity factor, %	77.5
Fixed charges	2.250
Maintenance	0.387
Operating	0.208
Steam	2.587
Total	5.432
	Annual cost, in dollars per Kw
Fixed charges	14.08
Maintenance	2.43
Operating	1.30

1. The need for increased profits to effect new exploration and field development and further expansion of present operating facilities.
2. The increasing of a geothermal power plant efficiency through a multipurpose plant utilizing all or many of the facets of a geothermal resource.

Costs of a geothermal power plant are dependent on many factors which man has no control over. A reservoir's pressure, temperature, and fluid characteristics were developed over millions of years. Various geological characteristics can govern the final output of the plant thus raising or lowering costs. Figure 69 shows the interrelation of the many and varied factors affecting a geothermal plant from inception to final power generation. Figure 70 shows the continuing effects of other factors on net power cost.

Actual dollar numbers are probably irrelevant due to our present large inflationary period and energy crisis unknowns. Dollars, and therefore required investments will change before this book goes to print. However, the presented criteria will define the feasibility of a geothermal investment. On the otherhand, if the energy crisis becomes acute we may find ourselves investing in geothermal power production with indifference to the costs.

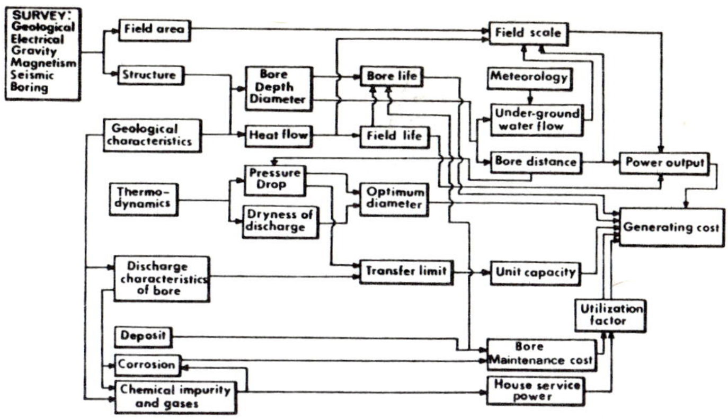

Figure 69. Many factors influence the cost of generating electricity.[122]

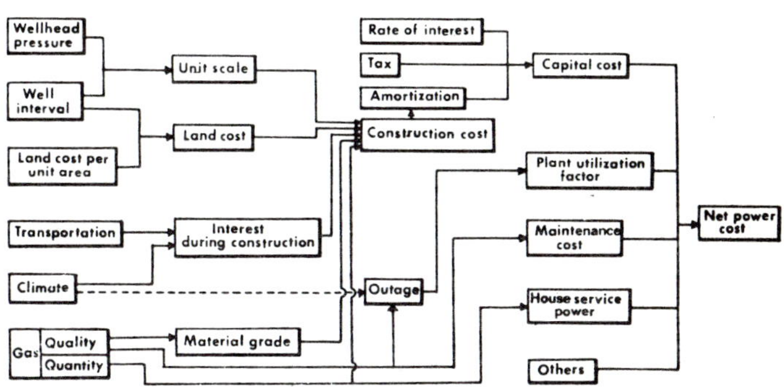

Figure 70. Other effects affecting geothermal power costs.[122]

CHAPTER 8

THE ENVIRONMENT

". . and the land was polluted . .Thus were they defiled with their own works, and went a whoring with their own inventions" . . . PSALM 106:38, 39

There will be an impact on the environment by geothermal energy. We must determine what this impact will be as this resource is developed and utilized under varying conditions. We must also determine the advantages and beneficial contributions to our society by - geothermal energy development and weight them against the disadvantages such as environmental defamation. However, let us not be taken aback because alarmists have spoken without fully examining all procedures, data, and the available technology for controlling pollution.

Claims espousing geothermal energy as the world's saviour as an abundant clean energy source are grossly misleading. Geothermal energy is a clean energy source from the standpoint that it is generated without a polluting effect unlike conventional fossil fuel or nucelar plants. Problems of pollution are inherent in the production of conventional energy because of the mobility of the raw product and the necessity to transform it into an unuseable byproduct to extract its heat.

Before we proceed to examine what type of environmental problems are derived from the development and utilization of geothermal energy, we will first note its relatively small impact on the environment including:

1. No strip mining;
2. No refining;
3. No industrial wastes to consider from processing (air, water, and solid);
4. No super tankers to collide or run aground;
5. No off-shore drilling;
6. No oil on the beaches and wildlife;
7. No hazard from fire as water in any form won't burn;
8. No need to transport energy;
9. No radioactive plant hazards;
10. No radioactive waste problems;
11. No thermal pollution of waters (if handled correctly);
12. A geothermal power plant uses less acres than a hydropower dam covers with water;
13. Unchecked sulfer emissions are much lower than a fossil fuel plant buring low sulfur fuel;

14. Consider Reyhaukur, Iceland: In the 1940's the skies were clouded with smoke as the city was barely visible – Today, after years of geothermal energy, the skies are beautiful, clear, and smokeless.

The list can go on and on, for the reader can think of many more energy related pollution problems. This is not to suggest geothermal resources don't pollute or affect the environment, but just to acknowledge pollution problems we are already encountering and that maybe through improved technology our energy needs could be satisfied with smaller detrimental effects.

Technology seems to be the underlying equation to many of our daily issues. To solve this equation is to solve many of our society's ills. Air and water treatment technologies are just today being examined and developed to the extent necessary to meet our pollution needs. Sampling techniques, data gathering and correlation, improved studies and impact statements, basic equipment and chemicals, etc., that were unheard of only a few years ago, are commonplace today. Therefore, we believe within ten years our pollution control technologies (i.e. air, water, wastewater, solid wastes) will be at such an improved state that they will be able to handle most of our current needs.

Although there are detrimental effects produced through the exploitation of a geothermal resource, we must be realists in a rather explosive issue. Any pollution and other problems associated with a geothermal resource can most probably be controlled with today's technology, and most definitely will be considered passe in the future.

Geothermal guidelines concerning geothermal power production, environmental impact statements and the leasing of federal lands for geothermal development have been set up.[116] Industry feels that too many environmental regulations may become inflexible and needlessly troublesome to the geothermal explorer. This type of situation is very unhealthy for it discourges the development and exploitation of the geothermal resource by the private sector. The geothermal industry's approach to the development of this resource involves (1) the relaxing of environmental zoning regulations to allow drillings to proceed and to be fully encouraged; (2) environmental studies and impact statements to be made after a useable geothermal resource is proven to exist; thus eliminating high cost studies for regions that may not be able to support geothermal energy; (3) allowing large tax incentives to geothermal developers by accepting tax-write offs for nonproducing wells; and, (4) the requirement of full disclosure of information by all developers to allow for a more rapid technology growth.

The disadvantage to an over zealous approach to geothermal energy development is the initial overlooking of many environmental problems which tend to build up and eventually must be taken care of at a cost probably many magnitudes greater than the initial investment would be. Then, with these large costs to be directly passed on to the public users of the energy or to be absorbed by the operating company, the great danger looming overhead becomes one of - whether regulations calling for such astronomical environmental costs are justified and legal. The hedging on requirements and guidelines and on compliance will proceed to run the whole concept of geothermal energy. That is, a relatively clean and alternate energy source to the conventional methods of today.

Land Pollution

The initial objections to the development of a geothermal field are usually based on questions of its impact on the surrounding environment. Many environmentalists and the general public have only within the past ten years become concerned with our seemingly lack of morality when it comes to the natural order and beauty of things. In fact, it is probably more an afterthought of the BIG TWO news promotors, the energy crisis and the pollution crisis. Environmentalists have won court cases across the country stopping construction that was detrimental to the aesthetics of the land. "Visual pollution" has resulted in a number of laws enacted to curb the continuous disruption of man's habitat. Implementation of high tension wires has had to go underground in some areas of the country and highways have had to go around other areas. Junk billboards and other eyesores have been legislated to be kept at a minimum.

Therefore, the aesthetic impact of the geothermal field is of utmost concern. Consider the Italians exploitation of their geothermal resource in the Larderello region. As of 1960 the localities of Larderello, Castelnuovo V.C., Sasso Pisano, Monterotaondo Moittino, Lago, Lagoni Rossi and Serrazano covered a geothermal reservoir which was tapped 357 times to an average depth of 500 meters over a total area of 45 Km^2. This results in a density of about 8 wells per square. In the Monte Amiata region 8 wells were drilled over an area of $3Km^2$. Across the small area of Travale 21 wells had been drilled by 1960 to an average depth of 570 m.[35]

The drilling for natural steam at the Geysers field, California, will reach 150 wells when the capacity of 1000 Mw is attained. This is to encompass an area of 12 miles.[123] This high density of wells per unit area is characteristic of the geothermal field development. From Figure 71, the reason for this becomes evident. This particular temperature survey was taken at a depth of one meter, but at much larger depths these temperatures could increase tenfold for a normal geothermal analomy. As the temperature range doubles its corresponding area decreases by two order of magnitudes. To double the temperature range again would reduce its representative area by another order of magnitude, the resultant area useable as a thermal resource is only a small fraction of the entire geothermal analomy.

In most geothermal regions not only are the drilling sites affected but the surrounding area must be set aside for future development. Development that would take place in the event of steam pressure reduction or increased plant requirements. Therefore, relatively large areas can be directly and indirectly affected by the exploitation of a geothermal resource.

Range (°C)	Area (m^2)
15-20	1,206,875
20-30	568,125
30-40	12,125
40-50	23,375
50-60	7,250
60-70	1,750

Figure 71. Typical areas of temperature zones of a geothermal analomy.[36]

The immediate geothermal region becomes a network of piping stretching for miles along the countryside as the steam from many wells is joined to operate one turbine. However, the detrimental effects to the aesthetics of a region by the piping can be kept at a minimum. The gathering lines occupy small strips of land that can be and in the case of Larderello, Italy, is, no more obscene than a network of irrigation canals. At Larderello agricultural interests are carried out among the steam gathering lines to the extent that farms, and fruit and vegetable production goes unhampered. Figure 72 shows enviro-geothermal exploitation at the Larderello field. Compare this with Figure 73 which is a canal irrigation network. Obviously, when man decides to be environmentally protective he can, for we see relatively little difference between the scars produced by the geothermal system and those produced by an irrigation system effecting lush crops.

Figure 72. The Larderello field, Italy, which has been envirogeothermally developed.

Figure 73. An irrigation canal network producing cash crops.[124]

The other entities of the geothermal production site are the wells and the power plant which also occupy small patches of the land. They need not be imposing

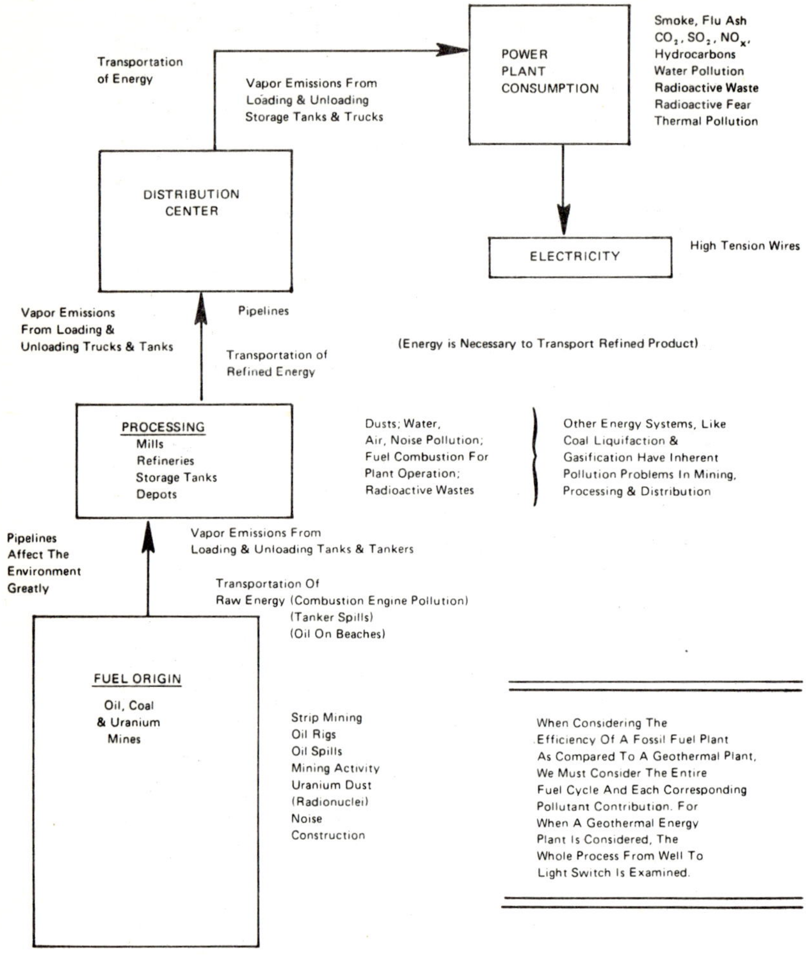

Figure 74. Conventional fuel cycle.

structures but can exist side by side with nature. At the Geysers the unused land is utilized for cattle grazing and a deer reserve.[123]

Probably the major environmental impact is observed before any energy is generated. Construction of roads into the wilderness, on-site construction of the power plants and pipelines, and the well drilling. The initial flurry of activity related to geothermal exploration can be burdensome for a region of beauty and solitude.

Noise from the construction, the drilling, the escaping of steam during testing is a disruptive impact that is equivalent to urban development. The decibel count is reduced dramatically to the level of an ordinary power plant after electricity is being generated. In the meantime, noise levels have been reduced through the use of mufflers and we believe present research is on the verge of a new technology to further decrease noise levels. For the present, technology has improved to the extent of substituting mud for air as the drilling circulatory fluid and special blow out preventive well heads have been designed (see Drilling Techniques). However, there are times when the wells must be allowed to blow steam such as when the steam reservoir is first penetrated or just before the well goes into production and the steam has to be cleansed of rocks and dust before entry into the turbines. During these periods extremely high noise levels are encountered which will have to be muffled in some manner.[123]

The geothermal resource is usually located in regions of lesser economic development and as previously mentioned these lands through geothermal development; i.e. water and irrigation, space heating and electricity, can become richly fertile and economically solvent. Furthermore, because these regions are sparsely populated, some detrimental environment effects associated with human needs (noise) can be programmed to meet desired levels.

A problem that could arise from the tapping of a geothermal resource is seismic disturbances. Although no geothermally induced effects have been observed at The Geysers over the past fifteen years nor at Larderello over the past 62 years, earth tremors induced by man's tampering with the subsurface unknown have had much publicity.

Wells dug for the purpose of waste disposal have resulted in earth tremors hundreds of miles from the well site as a result of pressure differences. Therefore, in the highly faulted geothermal structure already condusive to earthquake activity, the indiscriminate altering of the earth's delicate balance will eventually create problems of enormous magnitudes.

The releasing of high pressure steam and even relatively low pressure hot water may cause a shifting of the earth's crust along the series of fissures that have grown and become settled in place over millions of years. Furthermore, the very nature of the hot rock, dry geothermal resource; that is, to effect fissures under high pressure water that will continue to grow over the time period of usage, is the creation of a region quite similar to the subterranean network of cracks known in highly active fault areas. For these reasons we must take great pains to examine and re-examine the possible detrimental effects of geothermal resource exploitation. Much information can be obtained from the petroleum industry as their environmental studies with respect to piping, drilling, fluid removal, and other construction can be adapted to the geothermal resource. Special techniques concerning above ground piping (i.e. the Alaskan pipeline) might be utilized when relatively large piping distances for space heating become necessary. In any event, the long range environmental impact on land by the development of geothermal energy can not be known until it is operation. We can look at Larderello for optimistic results and encouragement to go forth and meet many of our energy needs with geothermal energy.

Water Impact

The water impact on the environment by the geothermal resource are many and varied. The initial problems associated with waters of geothermal origin is their high contaminate content. These include a variety of chemicals, high suspended solids levels, and high heat rejection. Note Table 31 - which is the analyses of geothermal waters from a number of Mexican wells. While the chemical content will vary from field to field, the general problems are the same.

Table 31. Geothermal Waters Content of some Mexican Wells.[103]

(pH about 7.5)

(Values in ppm)

Well No.	Na	K	Li	Ca	Mg	Cl	Br	I	Fe	SO_4	HCO_3	H_3BO_4	SiO_2	CO_2	H_2S	Total hardness (as $CaCO_3$)	Total Dissolved Solids (TDS)
1-A	4.450	600	12	210	30	7,420	5.2	1.0	nd	7.0	52	52	240	nd	nd	699	13.082
M-3	5.310	1.100	17	310	11	9.680	10.0	2.8	0.2	15.0	60	55	480	680	218	820	18.041
M-5 (-)	5.820	1.570	19	280	8	10.420	14.1	3.1	0.2	0.0	73	71	740	1.600	700	733	19.018
M-6	5.000	504	11	388	33	9.000	12.6	2.5	nd	16.4	158	21	151	420	37	1.106	18.412
M-7	5.250	910	13	230	18	9.310	9.2	2.6	nd	3.4	71	32	390	940	180	649	16.240
M-8 (-)	6.100	1.860	17	390	6	11.750	14.3	3.2	nd	0.0	890	115	770	nd	nd	1.000	21.915

The harmful effects of these contaminated waters on surface waters into which they are discharged have been seen at the Wairakei, New Zealand. The fish downstream of Wairakei plant have been affected by the plant's residual waters.[125] Furthermore, these geothermal waters exhibit highly corrosive, abrasive, and erosive characteristics and they must be treated before entry into the power plant turbines to prevent constant fouling. SiO_2, Na, K, Fe, are always found in geothermal fluids in appreciable amounts. Chlorides, sulfides, sulfates, NH_3, H_3BO_3, H_2S, and CO_2 in the waters contribute to the constant attack on steam power plant piping and other equipment.

Aside from the chemical aspects, geothermal water can become a major contributor to thermal pollution. Not only are the geothermal waters themselves a source of this pollution, but also the enormous quantities of cooling waters used in the power generation process and then discarded into a river, lake or ocean will lead to a degradation or at least altering of the surrounding environment. (for a schematic picture of this process see Chapter 5, Electrical Systems).

The problems associated with thermal and chemical pollution by geothermal waters are not unsolvable. At the Geysers, contaminated waters are reinjected into the ground (a method we previoulsy prescribed in other chapters). The purpose of this is twofold: (1) for the disposal of plant residuals and (2) to reestablish subterranean pressures to keep steam pressures constant and to prevent possible stresses and eventual seismic disturbances. As noted in the Land Impact subsection large water removal rates may cause subsidence. The geothermal reservoir at Wairakei is depleted by 70 million tons of water each year, obviously this much water will have an enourmous impact on the environemnt, by its existance and nonexistance.

Therefore, residual water reinjection is an important process for the alleviation of some geothermally induced environmental problems. However, depending on the nature of the geothermal reservoir and its waters composition, reinjection may not be practical or possible. Reinjected waters of high mineral content may cause the pores of reservoir rock to become filled with deposits from the relatively cooler processed water.

These waters can also be treated. Pre-power plant treatment must be accomplished at many geothermal energy producing sites. Dry steam from a geothermal well can usually be used as it leaves the ground. Wet steam is not so convenient to use. It usually contains many chemicals and minerals that must be removed before their entry into the power plant. This process is effected by numerous separators. Post plant treatment of the rejected geothermal waters must be incorporated in the geothermal plant system. Dry cooling towers might be utilized to cut down on thermal pollution by dissipating the leftover geothermal heat to the atmosphere and preventing the heat transfer to surface waters.

On the other hand, the answer seems simple enough: the complete utilization of the geothermal energy source, as discussed in previous chapters. After the initial steam is used to generate electricity, it is recombined with the waters from which it was separated before entering the steam turbines. These hot waters are then

1. treated and processed for the removal of chemicals and minerals. Boron, sulfur, diatomite and many other minerals and chemicals can be and are extracted from geothermal waters for a profitable operation; then,
2. these waters, having a high heat content can be used for space heating, agricultural, commercial, industrial and other hot water needs, or
3. passed through tertiery treatment to be used for daily consumption.

Thus, the chemical and thermal pollution can be prevented with many beneficial byproducts. This point cannot be stressed enough: that by itself a geothermal power plant can be highly inefficient and very pollutive; but through a step by step utilization of all the resources of a geothermal resource, pollution can be stymied and we can reap many benefits including fertile lands where deserts now lay.

Air Impact

When comparing the environmental impact of the geothermal source to that of the conventional fossil fuel and nuclear the result become ludicrous. The impact on air, land, and water of the geothermal resource is but a small fraction of the pollution spewn forth by our presently operating power plants. Aside from the actual plant site generation pollution, water, land and air are degraded through soil, oil and uranium mining. Petroleum shipping and receiving depots, petroleum vapors released during loading and unlaoding operations; crude oil processing pollutes both the air and water, strip mining for coal devastates the environment, oil rigs are more unsightly than than a few steam pipes running on the ground; the pollution from combustion of fuels to operate processing plants, oil tankers, railroads and trucks; wastes from processing uranium; including refinement, enrichment and conversion; are just a few of the numerous off-site pollution generating processes attributable to energy production which are not necessary for the generation of electricity by geothermal energy.

The actual on site consumption of conventional fuels, oil, coal and uranium may produce nitrogen oxides, fly ash, smoke, sulfur compounds including H_2S, CO_2, radioactive wastes other aerosols, thermal waters, this, the overall oil, coal, and/or uranium fuel cycles; beginning from their extraction from the earth, their transportation to milling and refinery sites, the milling and refining processes, the transportation to distribution centers, to the plant site, the consumption of the fuel, the resultant waste products, effect multiple sources of pollution which must be considered when the determination and ultimate comparison of geothermal energy systems pollution contributions are made.

The actual exhaust from a geothermal well is much different in nature and far less than that of other power plants. Nearly all, 99.5% of the exhaust is water vapor,[123] yet this percentage can vary from well to well. Noncondensable gases also vary percentagewise with each well, but the relative magnitudes of each constituent are just about the same. Carbon dioxide, 80–95%; hydrogen sulfide, 2–5%; hydrogen, 2–5%, methane, 2–5%, nitrogen, 1%, are the common constituents – making up the non-condensable gases or the remaineder of the exhaust stream.

However, even the relatively small amounts of hydrogen sulfide emitted by a geothermal power plant can be environmentally degrading. At Wairakei hydrogen sulfide emissions are in amounts large enough to corrode silverware and brass in nearby towns. The large, constant and widespread discharging of carbon dioxide into the atmosphere that will occur with the eventual expansion of geothermal exploitation is expected to have far reaching effects on world climatology. Large quantities of carbon dioxide in the atmosphere are thought to produce the "greenhouse effect", or the increasing of solar energy retention by the earth.[125]

With respect to a comparable sized fossil fuel plant, geothermal emissions are small. Comparing similar steam conditions and power outputs it is seen that sulfur dioxide emissions of geothermal plant are about one quarter those of a fossil fuel plant, or for a 1000 Mw power plant SO_2 emissions are as follows:

140 tons/day	fossil fuel (1% sulfur)
34 tons/day	geothermal energy

Furthermore, a new process has been developed[123] which catalizes the oxidation of hydrogen sulfide to sulfur and water and with the successful implementation of this process, it is expected to reduce sulfur dioxide emissions to even lower levels. Carbon dioxide emissions of a fossil fuel plant are nearly twenty times the geothermal plant. i.e. 20,000 tons/day to 860 tons/day for a comparable 1000 Mw plant.

Thus we note the major detrimental affects to the environemnt by a geothermal plant are effected by its construction. The actual plant producing pollutants are minimal and with extended processes can be controlled. In fact, many residual wastes can be turned into profitable enterprises. A major theme that must be emphasized is the simultaneous development of pollution technology along with geothermal technology. By the time we will be ready for geothermal power systems to go into operation, the pollution technology (of air, water, land, noise) will be ready to preserve the environment.

CHAPTER 9

THE FUTURE OF GEOTHERMAL ENERGY

"And He prayed again, and the heaven gave rain, and the earth brought forth her fruit" . . . . JAMES 5:18

Despite a few important areas of uncertainty, the future for geothermal power looks very bright. From the technical, economic, and resource potential standpoints, geothermal production of electricity has demonstrated itself worthy of considerable research and investigation. It is quite conceivable that this form of power production will at least augment the total power production of the western U.S. to the extent that hydropower does. A figure of 5 to 10% of the total western U.S. electrical power production has been suggested.

It is important at this early stage of the development of geothermal power production that extensive research into the possible problem areas of geothermal power be conducted before full-scale investment in production systems occurs. There must be positive evidence presented to the public that: (1) the geologic sources are not going to be short lived because of large withdrawals of fluid and heat, (2) seismic activity and ground subsidence will not increase to a dangerous level after geothermal fluid withdrawal or water injection, (3) air and water pollution from geothermal steam impurities will truly be minimal, (4) the withdrawal of geothermal fluids will not affect the freshwater table to any extent that might prove harmful to the environment, and (5) a standardized method of cost-analysis which includes all of the costs of geothermal power shows that it is competitive with conventional forms of electric power production.

It is important that the errors of the past not be repeated with geothermal energy by investing so heavily in the production machinery that it is difficult to stop if strong negative aspects (especially in the environmental area) are found.

Additional research is called for in the areas of low-temperature generating plants and multiple-use plants. Herein may lie considerable expansion of useful resources and important economic savings.

The high cost of development of geothermal resources may easily be beyond the reach of present electric utility companies. The oil industry, on the other hand, has the financial resources and, apparently, the interest. It has been reported recently that Standard Oil of New Jersey, Shell Oil, Continental Oil, and Union Oil have been making inquiries concerning geothermal land leases. It is highly questionable whether or not it is the best interest of the public to have geothermal power in the U.S. developed and hence owned by the oil industry.

Similarities between geothermal and oil-industry exploration and well drilling, as well as the propensity of the oil conglomerates to expand into nonpetroleum energy to the extent of control, strongly suggest that geothermal energy resources must be placed largely under federal control and investment. Despite the suspicion of collusion between industry and government to the exclusion of the public, it is imperative that the public have the control over geothermal resources and their development that federal operation affords.

Some reports readily conclude that geothermal power is completely safe and clean. This is far from certain at our present state of knowledge. Thus, it is absolutely imperative that those who (be they private or federal agencies) exploit geothermal resources be required to submit an environmental impact statement for public review prior to extensive development.

Environmental Impact of Geothermal Energy and the Future

To gain proper perspective on the environmental impact of producing electricity from a geothermal plant, it is necessary to understand the basic character of its various manifestations and to compare the relative impact of other thermal power cyclesnuclear and fossil fuel, since each produces its own effects. The kinds of effects produced may be categorized by their impact on the land, on the air, and on the water. Chapter 8 has discussed environmental problems; a further comment is somewhat in order for any look into the future.

Impact on the Land

The impact of the construction of wells, pipelines, and power plants is most evident during the development period, and for a large field this could extend over several years.

Drilling operations in a geothermal field have a considerable noise impact on the environment. The noise problem is associated mainly with drilling operations and steam escape during testing. The noise level usually drops to that of other power plants once the field is in production. Mufflers are now used at The Geysers field during drilling and testing operations and have significantly lowered the noise level.

Land subsidence is another potential effect of geothermal development. Subsidence can occur whenever support is removed from beneath the ground. It has been noted in oil fields, in mines, and from pumping of subsurface waters. When subsidence is caused by removal of ground water, then the field is usually located at shallow depths. It is now standard practice to reinject these wells with waste water.

Impact on the Air

Air pollution is also a significant problem, since noxious gases are often a by-product of geothermal wells. The gases produced are far less and is of different nature than that of a fossil fuel plant, the reason being that geothermal plants operate without combustion. There are many noncondensable gases which exist in geothermal steam. They include carbon dioxide, methane, hydrogen, nitrogen, ammonia, and hydrogen sulfude. Of these hydrogen sulfide presents the most serious environmental problems, even though it has the smallest release. Hydrogen sulfide dissolves readily in water and escapes into the atmosphere by evaporation.

Impact on the Waters

The natural thermodynamic constraints placed on any steam cycle require the reinjection of 60 to 70% of the total energy produced. This is normally done by circulating cooling waters through the condenser to pick up this reject heat and dissipate it into a larger body of water such as a river, lake, or ocean. Rejection of heat into the body of water can cause local environmental degradation, and is generally described as thermal pollution. It is present in all of the new types of power production methods being used. One way to alleviate thermal pollution is to reject the waste heat directly to the atmosphere and not through an intermediate body of water, as is now the practice. Cooling towers will accomplish this transfer, but require large quantities of low cost water. With the present system of production of electricity from the dry steam fields, all of this waste heat is either returned to the producing reservoir or reinjected directly to the atmosphere by the cooling towers, thus creating no thermal pollution.

GLOSSARY

Anticline – Rock Layers that are pushed together to form an upward bend in the shape of an arch.

Ash – Powdery bits of lava, also known as volcanic dust. Some is light, some dark, depending on its mineral composition.

Basalt – A dark, tough, fine-grained to dense, extrusive volcanic rock commonly occurring in sheetlike lava flows.

Bomb – A sizeable lump of lava, often in a very interesting shape. Bread-crust bombs have a cracked surface. Their curious appearance is due to their cooling and hardening on the outside while the interior was still hot, soft, and expanding.

Brine – A heavily saturated salt solution.

Cinders – Small bits of lava, but bigger than ash.

Cone – A heap of rock and lava around the outside of the volcano.

Continental Drift – The theory that continents slowly shift their positions as a result of currents in the molten rocks of the earth's mantle.

Crater – The scooped-out top of the volcano, generally high at the outside, getting deeper toward the center.

Crust – The solid, rocky, outer portion or shell of the earth; also known as the lithosphere.

Desalination – The removal of salt.

Diatomaceous Sediments – Earthy deposits formed mainly of the siliceous shells of diatoms; also known as diatomite.

Diastrophism – The process by which the earth's surface is reshaped through rock movements and displacements.

Dry Steam – Steam which has a very low vapor content.

Earthquake – An abrupt, violent movement of rocks along a fault.

Erosion – The process which constantly wears away rocks and carries off the crumbled bits to lower levels.

Fault – A fracture or zone of fractures in rock strata together with movement that displaces the sides relative to one another.

Fault Escarpment – In faults, upward shifts, produced by powerful upward pressures from below, pust up one side of the fault while the other side moves very little or remains at rest. This higher side of the fault is known as the "Fault Escarpment".

Fissure – A long, narrow, deep cleft or crack in the earth.

Fold – Folds are bends or wraps in rock formations.

Foreland – The area of a fold which is inclined and in the direction of the thrusting forces.

Fossil Fuel – Any natural fuel which is dug from the earth; coal, petroleum and natural gas are fossil fuels.

Fracture – The texture of the broken surface of a mineral as distinct from its cleavage.

Fumarole – A vent in a volcanic area from which smoke and gases arise.

Geochemist – A person who specializes in the branch of chemistry dealing with the chemical composition of the earth's crust and the chemical changes that occur there.

Geo Heat – Heat supplied by a geothermal resource.

Geothermal Resource – Any natural resource which will supply energy and heat; geysers, hot springs, boiling mud pools are geothermal resources.

Geyser – A spring from which columns of boiling water and steam rush into the air at intervals.

Hinterland – The region opposite the inclines and force directions of the foreland in a fold.

Holocene – Designating, or, of the recent epoch of geological time.

Hot Spot – A location where a geothermal resource is situated.

Hot Spring – A spring which is created when the quantity of underground water is too great to be turned into steam by the magma, the heated water flows on until it comes out of holes in the ground.

Hot-Water System – A type of geothermal resource which exists when water temperatures are usually below the boiling point at atmospheric pressures ranging from 50°C - 125°C. The water from these systems is similar to the ground and surface waters of surrounding regions and usually appear as hot springs.

Hydrology – The science dealing with the waters of the earth, their distribution on the surface and underground, and the cycle involving evaporation, precipitation, flow to the seas, etc.

Hydrothermal – Having to do with the action of hot water in producing materials and springs or in dissolving, shifting, and otherwise changing the distribution of minerals in the earth's crust.

Igneous Rock – Rocks produced by the action of fire; specifically formed by volcanic action or intense heat, as rocks solidified from molten magma at or below the surface of the earth.

Impervious Rock – Rocks incapable of being passed through or penetrated.

Lava – Magma that hardens after coming to the earth's surface.

Magma – Hot molten rock beneath the earth's crust.

Mantle – The layer of the earth's interior between the crust and the core.

Metamorphic Rocks – Sedimentary rocks that are raised from the sea bed, and undergo changes due to pressure and heat.

Obsidian – A hard, usually dark-colored or black, volcanic glass with conchoidal fracture. It is volcanic rock that cooled too rapidly to leave any time for the formation of crystals or coarse grains.

Orogenic Zones – Areas that were formed by structural disturbance of the earth's crust due to folding and faulting.

Potash – Any substance containing potassium.

Pumice – Pumice is obsidian that contained a lot of gas bubbles when it was hot. The bubbles formed holes that made this mineral very light.

Radiogenic – Produced by radioactivity.

Sedimentary Rocks – A group of rocks that is made of nothing more than debris, waste and rubble.

Seismology – A geophysical science dealing with earthquakes and related phenomena.

Steam Well – Fissure in ground capable of producing dry or wet steam.

Stratigraphy – The branch of geology dealing with the study of the nature, distribution, and relations of the stratified rocks of the earth's crust.

Tectonic – Pertaining to changes in the structure of the earth's crust, the forces responsible for such deformations and the external forms produced.

Topography – A study or description of a region showing all surface features.

Volcanic Mud – Ash mixed with water. Hot ash is forced high up into much cooler air. The ash condenses the moisture in the atmosphere, bringing on rain. When it comes down on the volcanic matter, it forms a thick mud.

Volcano – A vent in the earth's crust through which molten rock (lava), rock fragments, gases, ashes, etc. are ejected from the earth's interior. There are two types of volcanoes: an "active" volcano which is one that has erupted quite recently and that gives indications of eruption again. An "extinct" or "dead" volcano is one that has finished erupting.

Vulvanism – Volcanic activity or phenomena.

Wet Steam – Steam that has a high vapor content.

REFERENCES

1. "Conservation alone is not enough: the past – the future", *New York Times*, January 26, 1975.
2. P. Kruger, and C. Otte, *Geothermal Energy*, ed. P. Kruger, and C. Otte, (Stanford, California: Stanford University Press, 1973).
3. National Petroleum Council. "U.S. Energy Outlook: An Interim Report", November 1971.
4. "Around the World, Alternate Energy is Sought", *N.Y. Times*, pg. 23 (Jan. 26, 1975).
5. R. G. Bowen, "Potentially the West's Major Energy Resource", *Geothermal Energy*, Vol. 2, No. 12, December, 1974, pp. 29–30.
6. J. Barnea, "Geothermal Power", *Scientific American*, V. 226, January 1972, pp. 70–77.
7. J. J. O'Brien, "Geothermal Resources as a Source of Water Supply", *Journal of American Water Works Association*, V. 64, November, 1972, pp. 694–700.
8. R. Schuster, "Turning Turbines with Geothermal Steam", *Power Engineering*, V. 76, March 1972, pp. 36–41.
9. J. B. Koenig, "Worldwide Status of Geothermal Resources Development", *Geothermal Energy*, ed. P. Kruger, and C. Otte, (Stanford, California, Stanford University Press), 1973.
10. E. F. Wehlage, "Geothermal Energy's Potential for Heating and Cooling in Food Processing", *Geothermal Energy*, V. 2, No. 12, December 1974, pp. 7–14.
11. J. B. Koenig, "Geothermal Exploration in the Western United States", *Geothermics*, (Instituto Internazionale per le Ricerche Geotermiche, Pisa, Italy), 1971.
12. H. C. H. Armstead, "What is Geothermal Energy?", *Geothermal Energy*, ed. H. C. H. Armstead, (UNESCO, Paris, France), 1973, pp. 15–18.
13. J. P. Finney, F. J. Miller, and D. B. Mills, "Geothermal Power Project of Pacific Gas and Electric Company at The Geysers, California", *IEEE Trans Power Applications and Systems*, V. 92, January, 1973, pp. 108–115.
14. E. F. Wehlage, "Geothermal Power: Can it Help Solve the Energy Crisis?", *Machine Design*, V. 45, May 3, 1973, pp. 30–32.
15. R. G. Bowen, and E. A. Groh, "Geothermal Earth's Prinoidial Energy", *Technology Review*, Oct./Nov. 1971, p. 46.
16. G. R. Robson, "Geothermal Electricity Production", *Science,* V. 184, April 19, 1974, pp. 371–375.
17. D. E. White, "Characteristics of Geothermal Resources", *Geothermal Energy*, ed. P. Kruger, and C. Otte, (Stanford, California, Stanford University Press), 1973.

18. "Geothermal Energy: An Emerging Major Resource", *Science*, V. 177, September 15, 1972, pp. 978–980.
19. M. C. Smith, "The Los Alamos Geothermal Dry Hot Rock Source Demonstration Project", *Geothermal Energy*, V. 2, October, 1974, pp. 8–11.
20. W. O'Keefe, "Geothermal Power: Sleeping Giant Stirs, But Will Require Years to Waken Fully", *Power*, V. 117, April 1973, pp. 32–34.
21. "Geothermal Resources", *U.S. Geological Survey Research 1973*, U.S. Department of Interior, 1973, pp. 14–17.
22. "Geothermal Resources", *U.S. Geological Survey Research 1972*, U.S. Department of Interior, 1972, pp. 10–11.
23. J. Gilluly, A. C. Waters, and A. O. Woodford, *Principles of Geology*, (W. H. Freeman and Co., San Francisco), 1968.
24. *The Earth*, (Nelson Doubleday, United States of America), 1969.
25. E. Bullard, "Basic Theories", *Geothermal Energy*, ed. H. C. H. Armstead, (UNESCO, Paris, France), 1973, pp. 19–29.
26. G. Sestini, "Heat-Flow Measurement in Non-Homogeneous Terrains. Its Application to Geothermal Areas", *Geothermics*, (Instituto Internazionale per le Ricerche Geotesmiche, Pisa, Italy), V. 2, Part 1, pp. 424–436, 1970.
27. W. Sullivan, "The Restless Continents", *The New York Times Magazine*, January 12, 1975, p. 5.
28. S. Kavlakoglu, "Origin of Geothermal Waters or Natural Steam", *Geothermics*, (Instituto Internazionale per le Ricerche Geotesmiche, Pisa, Italy), V. 2, Part 2, 1970, pp. 1250–1255.
29. J. Barnea, "Economics of Multipurpose Use of Geothermal Resources", *Geothermal Energy*, V. 2, N. 11, November 1974, pp. 29–34.
30. L. R. Porter, "Geothermal Resource Investigations", *ASCE Journal of the Hydraulics Division,* V. 99, N. HY11, Nov., 1973, pp. 2097–2111.
31. P. Mathews, "Geothermal Experience at Geysers Power Plant", *ASCE Journal of the Power Division*, V. 99, N. PO2, Nov. 1973, pp. 329–338.
32. "Geothermal Dig In Montana", *Mechanical Engineering*, V. 96, September 1974, p. 45.
33. "Mile-Deep Well Drilled to Tap Geothermal Energy", *Machine Design*, V. 46, July 25, 1974, p. 12.
34. C. H. Jacoby, D. K. Paul, "Salt Domes as a Source of Geothermal Energy", *Mining Engineering*, Vol. 26, No. 5, May 1974, pp. 34–39.
35. R. Caltaldi, P. Ceron, P. DiMario, and T. Leardini, "Progress Report on Geothermal Development in Italy", *Geothermics*, (Instituto Internazionale per le Ricerche Geotemiche, Pisa, Italy), V. 2, Part 1, 77–87.
36. C. J. Banwell, and R. Gomez Valle, "Geothermal Exploration in Mexico 1968–1969", *Geothermics*, (Instituto Internazionale per le Ricerche Geotemiche, Pisa, Italy), V. 2, Part 1.
37. "New Geothermal Field Starts Producing", *Electrical World*, August 1, 1972, pp. 26–29.

38. K. Sato, "The Present State of Geothermal Development in Japan", *Geothermics*, (Instituto Internazionale per le Ricerche Geotemiche, Pisa, Italy), V. 2, Part 1, pp. 155–184.
39. S. Komagata, H. Iga, H. Nakamura, Y. Minohara, "The Status of Geothermal Utilization in Japan", *Geothermics*, (Instituto Internazionale per le Ricerche Geotemiche, Pisa, Italy), pp. 185–196.
40. A. N. Tikhonov, and I. M. Dvorov, "Development of Research and Utilization of Geothermal Resources in the U.S.S.R.", *Geothermics*, (Instituto Internazionale per le Ricerche Geotemiche, Pisa, Italy), pp. 1072–1078.
41. J. H. Smith, "Geothermal Development in New Zealand", *Geothermics*, (Instituto Internazionale per le Ricerche Geotemiche, Pisa, Italy), pp. 232–247.
42. W. A. Mahon, and J. B. Finlayson, "The Chemistry of the Broadlands Geothermal Area New Zealand", *American Journal of Science*, V. 272, January 1972, pp. 48–68.
43. "Ethiopia's Geothermal Possibilities", *Compressed Air Magazine*, Vol. 77, January 1972, pp. 14–16.
44. V. T. Radia, "Geothermal Energy Prospects in South Sulawesi, Indonesia", *Geothermics*, (Instituto Internazionale per le Ricerche Geotemiche, Pisa, Italy), pp. 136–149.
45. A. Ten Dam, C. Erentoz, "Kizildere Geothermal Field-Western Amatolia", *Geothermics*, (Instituto Internazionale per le Ricerche Geotemiche, Pisa, Italy), pp. 124–129.
46. L. T. Braun, "Geothermal Exploration In Nicaragua", *Geothermics*, (Instituto Internazionale per le Ricerche Geotemiche, Pisa, Italy), pp. 41–42.
47. W. T. Cheng, "Geophysical Exploration in the Tatun Volcanic Region, Taiwan", *Geothermics*, (Instituto Internazionale per le Ricerche Geotemiche, Pisa, Italy), pp. 262–274.
48. E. E. Arango, A. J. Burtrago, R. Cataldi, G. C. Ferrar, C. Panichi, V. J. Villegas, "Preliminary Study on the Ruiz Geothermal Project (Columbia), *Geothermics*, (Instituto Internazionale per le Ricerche Geotemiche, Pisa, Italy), pp. 43–56.
49. R. Fernandez-Rivas, "Geothermal Resources of Guatemala, Central America", *Geothermics*, (Instituto Internazionale per le Ricerche Geotemiche, Pisa, Italy), pp. 1015–1025.
50. "Geothermal Resources of Costa Rica", *Geothermics*, (Instituto Internazionale per le Ricerche Geotemiche, Pisa, Italy), pp. 1026–1029.
51. C. G. Dixon, W. H. Morton, "Thermal and Mineral Springs in Uganda", *Geothermics*, (Instituto Internazionale per le Ricerche Geotemiche, Pisa, Italy), pp. 1035–1038.
52. M. A. Nzaro, "Geothermal Resources of Tanzania", *Geothermics*, (Instituto Internazionale per le Ricerche Geotemiche, Pisa, Italy), pp. 1039–1043.
53. B. R. R. Iyengas, "Geothermal Areas of Czechoslovakia", *Geothermics*, (Instituto Internazionale per le Ricerche Geotemiche, Pisa, Italy), pp. 1044–1049.

54. S. Klir, "Geothermal Areas of Czechoslovakia", *Geothermics*, (Instituto Internazionale per le Ricerche Geotemiche, Pisa, Italy), pp. 1055–1058.
55. B. Kremar, B. M. Lanovic, "Geothermal Exploration of Hot Water Sources in the Carpathians of Yugoslavia and Czechoslavakia", *Geothermics*, (Instituto Internazionale per le Ricerche Geotemiche, Pisa, Italy), pp. 1059–1062.
56. T. Boldizsar, "Geothermal Energy Production from Porous Sediments in Hungary", *Geothermics*, (Instituto Internazionale per le Ricerche Geotemiche, Pisa, Italy), pp. 99–109.
57. J. Dowgiallo, "Occurrence and Utilization of Thermal Waters in Poland", *Geothermics*, (Instituto Internazionale per le Ricerche Geotemiche, Pisa, Italy), pp. 95–98.
58. S. Bjornsson, "A Program for the Exploration of High Temperature Areas in Iceland", *Geothermics*, (Instituto Internazionale per le Ricerche Geotemiche, Pisa, Italy), pp. 1050–1054.
59. J. R. McNitt, "The Role of Geology and Hydrology in Geothermal Exploration", *Geothermal Energy*, ed. H. C. H. Armstead, (UNESCO, Paris, France), 1973, pp. 33–40.
60. H. G. Wunderlich, "Geothermal Resources and Present Orogenic Activity", *Geothermics*, (Instituto Internazionale per le Ricerche Geotemiche, Pisa, Italy), pp. 1226–1230.
61. H. Nakamura, K. Sumi, K. Katagiri, T. Iwata, "The Geological Environemnt of Matsukawa Geothermal Area, Japan", *Geothermics*, (Instituto Internazionale per le Ricerche Geotemiche, Pisa, Italy), pp. 221–231.
62. D. W. Strangeway, "Geophysical Exploration Through Geologic Cover", *Geothermics*, (Instituto Internazionale per le Ricerche Geotemiche, Pisa, Italy), pp. 1231–1234.
63. M. P. Hochstein, and T. M. Hunt, "Seismic, Gravity and Magnetic Studies, Broadlands Geothermal Field, New Zealand", *Geothermics*, (Instituto Internazionale per le Richerche Geotemiche, Pisa, Italy), pp. 333–346.
64. T. Meidov, "Application of Electrical Resistivity and Gravimetry in Deep Geothermal Exploration", *Geothermics*, (Instituto Internazionale per le Ricerche Geotemiche, Pisa, Italy), pp. 303–310.
65. G. W. Grindley, "Subsurface Structures and Relation to Steam Production in the Broadlands Geothermal Field, New Zealand", *Geothermics*, (Instituto Internazionale per le Ricerche Geotemiche, Pisa, Italy), pp. 248.
66. E. J. Douze, and G. G. Sorrells, "Geothermal Ground-Noise Surveys", *Geophysics*, V. 37, N. 5, October 1972, pp. 813–824.
67. G. V. Keller, "Induction Methods in Prospecting for Hot Water", *Geothermics*, (Instituto Internazionale per le Ricerche Geotemiche, Pisa, Italy), pp. 318–332.
68. J. J. Jacobson, "Deep Electromagnetic Sounding Technique", *D. Sc. Thesis T-1252*, Colorado School of Mining, 1969.

69. G. F. Risk, W. J. P. MacDonald, G. B. Dawson, "D.C. Resistivity Surveys of the Broadlands Geothermal Region, New Zealand", *Geothermics*, (Instituto Internazionale per le Ricerche Geotemiche, Pisa, Italy), pp. 287–302.
70. E. J. Moore, "A Graphical Description of a New Method for Determining Equivalent NaCl Concentration from Chemical Analyses", (Proceedings 7th Annual Well Logging Symposium, Society Professional Well Log Analysts, Houston, Texas), 1966.
71. G. Palmason, J. D. Friedman, R. S. Williams, Jr., J. Jonsson, K. Salmundsson, "Aerial Infrared Surveys of Reykjanes and Torfajokull Thermal Areas, Iceland", *Geothermics*, (Instituto Internazionale per le Ricerche Geotemiche, Pisa, Italy), pp. 399–419.
72. D. T. Hodder, "Application of Remote Sensing to Geothermal Prospecting", *Geothermics*, (Instituto Internazionale per le Ricerche Geotemiche, Pisa, Italy), pp. 368–380.
73. R. Cassini, C. M. Maimo, A. M. Torelli, "Ground and Airborne Thermal Imagery on Italian Volcanic Areas", *Geothermics*, (Instituto Internazionale per le Ricerche Geotemiche, Pisa, Italy), pp. 413–410.
74. F. Tonani, "Geochemical Methods of Exploration for Geothermal Energy", *Geothermics*, (Instituto Internazionale per le Ricerche Geotemiche, Pisa, Italy), pp. 492–515.
75. A. J. Ellis, "Quantitative Interpretation of Chemical Characteristics of Hydrothermal Systems, *Geothermics*, (Instituto Internazionale per le Ricerche Geotemiche, Pisa, Italy), pp. 516–528.
76. G. E. Siqualdason, "Geothermal Methods in Geothermal Exploration", ed. H. C. H. Armstead, (UNESCO, Paris, France), 1973, pp. 49–59.
77. A. J. Ellis, and W. A. Mahon, "Natural Hydrothermal Systems and Hot Water /Rock Interactions", Geochim, et. cosmoch, Acta, V. 28, 1964, pp. 1323–1357.
78. W. B. Stilwell, "Drilling Practices and Equipment in Use at Wairakei", *Geothermics*, (Instituto Internazionale per le Ricerche Geotemiche, Pisa, Italy), pp. 714–724.
79. U. Cigni, "Machinery and Equipment for Harnessing of Endogenous Fluid", *Geothermics*, (Instituto Internazionale per le Ricerche Geotemiche, Pisa, Italy), pp. 704–713.
80. K. Matsuo, "Drilling for Geothermal Steam and Hotwater", *Geothermal Energy*, ed. H. C. H. Armstead, (UNESCO, Paris, France), 1973, pp. 73–83.
81. K. Matsuo, "Present State of Drilling and Repairing of Geothermal Production Wells in Japan", *Geothermics*, (Instituto Internazionale per le Ricerche Geotemiche, Pisa, Italy), pp. 1467–1479.
82. F. Fabbri, and M. Vidali, "Drilling Mud in Geothermal Wells", *Geothermics*, (Instituto Internazionale per le Ricerche Geotemiche, Pisa, Italy), pp. 735–741.

83. J. Cromling, "Geothermal Drilling in California", *Journal of Petroleum Technology*, V. 25, September 1973, pp. 1033–1038.
84. Y. Nakajima, "Geothermal Drilling in the Matsukawa Area", *Geothermics*, (Instituto Internazionale per le Ricerche Geotemiche, Pisa, Italy), pp. 1480–1496.
85. U. Cigni, A. Giovanni, E. Luschi, M. Vidali, "Completion of Producing Geothermal Wells", *Geothermics*, (Instituto Internazaionale per le Ricerche Geotemiche, Pisa, Italy), pp. 757–763.
86. *Geothermal Energy*, V. 2, No. 1, November 1974, p. 64.
87. L. E. Garrison, "Geothermal Steam in The Geysers – Clear Lake Region, California", *Bulletin of Geological Society of America*, V. 83, May 1972, pp. 1449–1468.
88. E. T. Allen, and A. L. Day, "Steam Wells and Other Thermal Activity at 'The Geysers', California", *Carnegie Institute Washington Publication 378*, 1927, pp. 1–106.
89. J. R. McIvitt, "Exploration and Development of Geothermal Power in California", *California Division of Mines and Geology, Spec. Dept. 75*, 1963.
90. D. E. White, "Thermal Waters of Volcanic Origin", *Bulletin of Geological Society of America*, V. 68, No. 12, Dec., 1957, pp. 1637–1658.
91. D. E. White, L. J. P. Muffler, and H. A. Truesdell, "Vapor-dominated Hydrothermal Systems Compared with Hot Water Systems", Econ Geology, V. 66, No. 1, Jan. 1971, pp. 75–97.
92. H. Sato, "On Matsukawa Geothermal Power Plant", *Geothermics*, (Instituto Internazionale per le Ricerche Geotemiche, Pisa, Italy), pp. 1546–1551.
93. J. P. Finney, "The Geysers Geothermal Power Plant", *Chemical Engineering Progress*, V. 68, No. 7, July 1972, pp. 83–86.
94. A. Dal Secco, "Turbocompressors for Geothermal Plants", *Geothermics*, (Instituto Internazionale per le Ricerche Geotemiche, Pisa, Italy), pp. 819–833.
95. I. Ciapica, "Present Development of Turbines for Geothermal Application", *Geothermics*, (Instituto Internazionale per le Ricerche Geotemiche, Pisa, Italy), pp. 834–838.
96. G. Ricci, Ge. Viviani, "Maintenance Operations in Geothermal Power Plants", *Geothermics*, (Instituto Internazionale per le Ricerche Geotemiche, Pisa, Italy), pp. 839–847.
97. D. M. Wigley, "Recovery of Flash Steam from Hot Bore Water", *Geothermics*, (Instituto Internazionale per le Ricerche Geotemiche, Pisa, Italy), pp. 1588–1591.
98. G. Weismantle, "Geothermal Power Still Iffy", *Chemical Engineering*, March 5, 1973, pp. 40–41.
99. J. H. Anderson, "A Vapor Turbine Geothermal Power Plant", *Geothermics*, (Instituto Internazionale per le Ricerche Geotemiche, Pisa, Italy), pp. 1530–1532.

100. M. C. Smith, R. Potter, D. Brown and R. L. Alamodt, "Induction and Growth of Fractures in Hot Rock", *Geothermal Energy*, ed. P. Kruger, and C. Otte, (Stanford University Press, Stanford, California), 1973, pp. 251–268.
101. "Dry Geothermal Wells: Promising Experimental Results", *Science*, V. 182, October 1973, pp. 43–44.
102. A. D. K. Laird, "Water for Geothermal Resources", *Geothermal Energy*, ed. P. Kruger, and C. Otte, (Stanford University Press, Stanford, California), 1973, pp. 177–196.
103. C. M. Wong, "Geothermal Energy and Desalination: Partners in Progress", *Geothermics*, (Instituto Internazionale per le Ricerche Geotemiche, Pisa, Italy), pp. 892–895.
104. W. B. Murray, "Fresh Water from Brine: Part II", *Water and Sewage Works*, V. 119, Feb., 1972, pp. 54–57.
105. S. S. Einarsson, "Geothermal District Heating", *Geothermal Energy*, ed. H. C. H. Armstead, (UNESCO, Paris, France), 1973, pp. 15–18.
106. G. Palmason, and J. Zoega, "Geothermal Energy Developments in Iceland 1960–1969", *Geothermics*, (Instituto Internazionale per le Ricerche Geotemiche, Pisa, Italy), pp. 73–76.
107. F. A. Makarenko, B. F. Mavritsky, B. A. Lokshin, and V. I. Kononov, Geothermal Resources of the U.S.S.R. and Prospects for their Practical Use", *Geothermics*, (Instituto Internazionale per le Ricerche Geotemiche, Pisa, Italy), pp. 1086–1091.
108. G. Dragone, and O. Rumi, "Pilot Greenhouse for the Utilization of Low-Temperature Waters", *Geothermics*, (Instituto Internazionale per le Ricerche Geotemiche, Pisa, Italy), pp. 918–920.
109. S. H. Ross, "Geothermal Potential of Idaho", *Geothermics*, (Instituto Internazionale per le Ricerche Geotemiche, Pisa, Italy), pp. 975–1008.
110. G. Reynolds, "Coding with Geothermal Heat", *Geothermics*, (Instituto Internazionale per le Ricerche Geotemiche, Pisa, Italy), pp. 1658–1661.
111. B. Lindall, "Industrial and Other Applications of Geothermal Energy", *Geothermal Energy*, ed. by H. C. H. Armstead, (UNESCO, Paris, France), 1973, pp. 135–148.
112. J. H. Smith, "Geothermal Development in New Zealand", *Geothermics*, (Instituto Internazionale per le Ricerche Geotemiche, Pisa, Italy), pp. 232–274.
113. B. Lindall, "The Use of Natural Steam in a Diatomite Plant", *Geothermics*, (Instituto Internazionale per le Ricerche Geotemiche, Pisa, Italy), pp. 921–924.
114. "Geothermal Chemicals, *Chemical Engineering*, V. 81, April 29, 1974, p. 164.
115. B. Lindal, "The Production of Chemicals from Brine and Seawater Using Geothermal Energy", *Geothermics*, (Instituto Internazionale per le Ricerche Geometiche, Pisa, Italy), pp. 910–917.
116. "Critics Hit Geothermal Guidelines", *Electric World*, V. 176, Dec. 15, 1971, p. 32.

117. J. P. Finney, "Design and Operation of the Geysers Power Plant", ed. P. Kruger, and C. Otte, (Stanford University Press, Stanford, California), 1973, pp. 145–161.
118. R. Corti, P. Di Mario, F. Mondolfi, "New Trends in the Planning and Design of Geothermal Power Plants", *Geothermics*, (Instituto Internazionale per le Ricerche Geometiche, Pisa, Italy), pp. 768–779.
119. B. Wood, "Geothermal Power" *Geothermal Energy*, ed. H. C. H. Armstead, (UNESCO, Paris, France), 1973, pp. 109–121.
120. G. Pollaski, "Design and Construction of Steam Pipelines", *Geothermics*, (Instituto Internazionale per le Ricerche Geotemiche, Pisa, Italy), pp. 780–818.
121. J. H. Smith, "Collection and Transmission of Geothermal Fluids", ed. H. C. H. Armstead, (UNESCO, Paris, France), 1973, pp. 97–106.
122. T. Hayashida, "Cost Analysis on the Geothermal Power", *Geothermics*, (Instituto Internazionale per le Ricerche Geometiche, Pisa, Italy), pp. 950–957.
123. R. G. Bowen, "Environmental Impact of Geothermal Development", *Geothermal Energy*, ed. P. Kruger, and C. Otte, (Stanford University Press, Stanford, California), 1973, pp. 197–215.
124. A. C. Fisher, Jr., "Rhodesia, a House Divided", *National Geographic*, V. 147, No. 5, May 1975, p. 664.
125. W. Sullivan, "Analyst Warns of Pollution from Geothermal Projects", *New York Times*, March 1, 1975.

INDEX

A

B

C

D

E

F

G

H

I

J

K

L

M

N

O

P

R

S

T

U

V

W

Y